THE ULTIMATE
BIRDFEEDER
HANDBOOK

THE ULTIMATE
BIRDFEEDER
HANDBOOK

JOHN A. BURTON
Photography by STEVE YOUNG

Connaught

This edition first published in MMVII by New Holland Publishers (UK) Ltd
London • Cape Town • Sydney • Auckland
www.newhollandpublishers.com

Garfield House, 86–88 Edgware Road, London W2 2EA, United Kingdom

80 McKenzie Street, Cape Town 8001, South Africa

14 Aquatic Drive, Frenchs Forest, NSW 2086, Australia

218 Lake Road, Northcote, Auckland, New Zealand

ISBN-10: 1 84517 131 4
ISBN-13: 978 1 84517 131 5

Publishing Manager: Jo Hemmings
Project Editor: Camilla MacWhannell, Gareth Jones
Cover Design and Design concept: Alan Marshall
Designer: Shane O'Dwyer, D&N Publishing
Editor: Sylvia Sullivan
Index: Angie Hipkin
Production: Joan Woodroffe

Reproduction by Modern Age Repro Co., Hong Kong
Printed and bound in Malaysia by Times Offset (M) Sdn Bhd

Contents

Introduction 8

Section 5

Directory of Bird Foods 84

Introduction

I have been feeding birds for as long as I can remember. As a small boy growing up in post-war Britain, at first it was only ever with household scraps; by the time I was a teenager, I was putting out birdseed and peanuts. I lived in the suburbs of London, but close to a public woodland, and I can remember my excitement when I saw my first Nuthatch on the bird table. As a teenager, I became an avid birdwatcher and joined a ringing group, so that soon after my 16th birthday I applied for, and got, my ringing permit. I quickly learned which foods attracted which birds and I saw a wide range of the more common species. It is the data drawn from enthusiastic amateurs such as myself that has allowed the researchers at the British Trust for Ornithology to understand in detail about the breeding successes of various species, and also to monitor long-term changes. Among the most exciting personal examples was a Song Thrush I had ringed one summer in my London garden, which was later found (shot, unfortunately) in Portugal.

These and other experiences nearly half a century ago, led to a life-long interest in wildlife, which was also to develop, in due course, into an interest in conservation. In the 1950s, most species of birds were still reasonably abundant (Red-backed Shrikes even bred on a common in London's suburbs), while, in winter, massive flocks, including Bramblings and Tree Sparrows, gathered. Then, in the 1960s, catastrophic declines of birds of prey, due to pesticides, were noticed, and then other species declined too. From then on it seems there has been one problem after another for birds. Not everything is gloom and doom but, nonetheless, birds and other wildlife are threatened on a scale that is incomparable with anything in past centuries. Large parts of the countryside are now monoculture agriculture, with little or no space for wildlife, and the countryside is criss-crossed with high tension cables, with roads and street lighting all pervasive; even the recent rise of wind farms has created a significant threat to birds. No one growing up today can experience the numbers and abundance of birds that were found comparatively recently in the British countryside. But public awareness is greater than ever, and so we can expect positive action in the future.

With so few of the original habitats left in Britain, suburban gardens have, over the past half century, come to represent one of the most bio-diverse habitats. That is to say, within an area of a few acres of combined gardens, there will probably be more species of plants and animals than most other habitats in Britain. Not all these species are native to the UK but, nonetheless, even exotic garden flowers can help sustain populations of bees, Hoverflies, beetles and myriad other insects. As a consequence, birds not only find places to nest and breed but also find abundant food supplies; and when these are augmented at bird feeders, it can mean the difference between survival and death in winter.

There are numerous books on feeding birds and attracting wildlife, but surprisingly this volume is probably the first to address the nutritional aspects of bird feeding. While humans appear to be obsessed with counting calories and stuffing themselves with vitamins, until recently we have been content to simply string up peanuts, and throw out bread and handfuls of bird seed. Some of the seed suppliers started drawing attention to the nutritional benefits of certain types of sunflower and other seeds, and this stimulated me to look at all the foods and see which gave the greatest benefit to birds. It is an interesting area for research and one that the back garden birdwatcher can easily pursue. At its simplest level it is a case of putting out bowls of different seeds and observing the preferences of various species. Carried out over several years, such data can become invaluable.

I hope that by feeding birds, the reader will not only share some of the pleasures that I have enjoyed over the years but also, ultimately, be stimulated to take action in some way to help conserve what is left of our wildlife, and wildlife in other parts of the world. A starting point is to join a Wildlife Trust or one of the other organisations listed on page 123.

John Burton

Bird-feeding Basics

In the past, bird feeding was largely a means of trapping them for the pot, but for more than 100 years, we have been feeding them for the pleasure of seeing them in our garden, and bird feeding is now a multi-million pound business.

ABOVE: *In autumn, Jays wander far and wide in search of acorns, some of which they hide, thus helping disperse Oaks.* **OPPOSITE:** *Modern bird feeders are designed to make the birds, such as the Great and Blue Tits seen here, easily visible, keep food dry, and be easy to keep clean.*

Why feed birds?

Once upon a time, wild birds were generally fed as a means of trapping them for the cooking pot or to put them in cages. More recently, however, the fashion for keeping wild birds in captivity has declined, almost to the point of extinction, and so has the trapping of birds for the pot, at least in Britain. Parallel with this decline, there has been an increasing interest in watching living birds, and now millions of people get an enormous amount of pleasure from birdwatching, whether it be in the bleak marshes of north Norfolk or from the comfort of their living room window.

Feeding birds is an essential part of birdwatching: from peanuts in suburban gardens for Blue Tits to wheelbarrow loads of wheat for wild ducks and swans at a Wildfowl and Wetlands Trust centre; from tubes of sugar water for hummingbirds in California to dead cows and sheep for vultures in the Balkans. It's only a question of scale and location – just as a garden can vary in size from a few square metres to a huge nature reserve such as the RSPB's Minsmere in Suffolk. And the latter is every bit a 'garden', in as much as it is almost entirely man-made and is constantly being managed. In this respect it is no different from my own 'garden', which is also managed to benefit wildlife, albeit on a smaller scale, with a greater use of exotic flowering plants. This is a very important point that cannot be overemphasized. It is all a matter of size, and while it is certainly true that size matters, it is because of their sheer number that gardens now make up one of the largest and most significant habitats in Britain. Always remember that, just because you recognize your garden as finite, birds do not. As far as they are concerned, your garden is just one part of a vast habitat. You may only have one or two trees, but taken together with

BELOW: *To many people an allotment or vegetable patch is an essential part of their garden, and while some birds may eat young plants in particular, there is no doubt that on balance birds are both beneficial to, and benefit from, a well managed allotment.*

same designs as are still in use – there is a chapter on 'Feeding Birds in Winter', where Baron Von Berlepsch's famous bird food recipe is given in full. Berlepsch makes one point that is not often found in more recent books: '[F]ood must always be accessible to the birds, especially in sudden changes of weather, blizzards, wind, rain and frost, and must always be in the best condition.' I agree; it is no use feeding birds, and then when the weather turns nasty, forgetting to fill up the feeders – this is when they need it most of all.

Apart from making them more accessible and easier to see, feeding birds plays an increasing part in

ABOVE: *Fruit is high in sugars and carbohydrates, giving birds much-needed energy in winter. Although many species of birds, such as this Blue Tit, will readily feed on fruit that has fallen to the ground, some prefer the safety of it above ground level.*

those of your neighbours, they may constitute part of a very mixed 'woodland' of several hectares.

It is not much more than a century since bird feeding as we know it started. One of the first books on the subject, *How to Attract and Protect Wild Birds* by Martin Heisemann (a translation of *Beschreibung v. Berlepschscher Nisthölen*, by Baron Von Berlepsch) was published in 1907. The eleventh Duchess of Bedford wrote in the introduction: 'The following pages form an admirable treatise not only on bird protection, but more than that, on bird preservation. Game birds have long been carefully preserved… The Baron Von Berlepsch has applied the system of game management to all useful birds…' Although most of the book deals with creating nestboxes – essentially the

ABOVE: *One of the principal reasons for using feeders, such as this one, which has attracted a Great Tit and Nuthatch, is to make the birds more visible, so that the person feeding them gets pleasure as well. A good feeder should be attractive, easy to clean and allow the birds easy access to the food.*

their conservation. As I write, Bill Oddie is presenting his latest television series on birdwatching, and articles in the *Radio Times* have pointed out that suburban gardens are now among some of the most important habitats for several species of birds that are declining in Britain. I live in the middle of the East Anglian countryside – an area famed for some of Britain's finest nature reserves yet, at the same time, equally infamous for some of the most barren and wildlife unfriendly agricultural landscapes. There are thousands and thousands of hectares containing less biodiversity than a supermarket car park. As I drive thorough these monocultures of oilseed rape, wheat, barley and sugar beet, I wonder how birds can possibly survive in those weed-free fields? And they are not just weed free – the numbers of invertebrates, such as worms, slugs and insects and their larvae, are also dramatically reduced, often to the point of near extinction. It has long been recognized that

hedgerows, sadly also now much diminished in numbers, are important habitats for birds and other wildlife – essentially mimicking the rich habitat of forest edges and clearings. Suburban gardens create a very similar type of habitat: the lawns, shrubberies and flower-filled borders imitate the structure of a woodland glade.

Feeding birds in gardens, particularly during the lean winter months, helps many species to survive; they might otherwise have considerable difficulty in finding food. Feeding is, of course, only part of the solution and, even in a garden, it is important to ensure that some of the issues relevant to the wider countryside are addressed as well. Pollution, water, herbicides, pesticides and predators are just a few of the other issues that need considering.

A good wildlife garden will be sympathetically planted. By this I mean the locality, soil type, drainage, amount of sunlight and so forth must all be taken into account. In addition, it is perhaps obvious that the use of chemicals should be kept to an absolute minimum.

Of course, I have not yet mentioned that one of the fundamental reasons that most people feed birds is an aesthetic one – we simply like seeing birds, and we like to see as wide a variety as possible. It is these objectives that have the greatest influence on the way we feed.

BELOW: *In larger gardens rabbits and ants can often be a problem on lawns. Ants attract Green Woodpeckers, so do have benefits, and rabbits are relatively easy to exclude with wire netting, with the bottom bent outwards to stop the rabbits burrowing under it.*

LEFT: *Greenfinches are among the commonest birds at feeders, and are particularly fond of peanuts. However, they often occur in large numbers and are aggressive towards other birds. If you present a range of feeders, other birds will not be discouraged by the Greenfinches.*

ABOVE: *The Woodpigeon colonized towns and cities during the 20th century and is one of the largest birds to be found regularly at feeding stations. Although they are greedy feeders, they are not agile, so cannot deal with hanging feeders.*

ABOVE: *A classic bird table with suspended feeders, showing the range of birds that might be attracted to a suburban garden almost anywhere in Britain. The birds include a Robin, Nuthatch, Coal Tit, Great Spotted Woodpecker, Woodpigeon Long-tailed Tits, Goldfinch, Blue Tit and Greenfinches.*

Setting up a basic feeding station

When setting up a feeding station the three main considerations are, to quote the estate agents' adage, 'location, location and location'. The geographical location will determine which species are likely to occur in the general area; the location within that area – the size and type of garden – will have an even more important influence, and the location of feeders and food within the garden itself will be important in attracting birds and ensuring they are not snatched by predators. A garden in the Scottish Highlands, for instance, is one of the few places in the British Isles where it is reasonable to expect a Crested Tit; the West of England and Wales are places where Pied Flycatchers can occur, and Ring-necked Parakeets are currently not found in many areas away from London. A large garden with ponds may be visited by herons, Mallards and even possibly a Kingfisher – all unlikely to turn up in a small garden behind urban terraces. Chaffinches prefer to feed on the ground, while Great Spotted Woodpeckers prefer to be off the ground.

Assuming that a typical garden consists of flowerbeds, shrubs, perhaps a vegetable patch, a few trees and a lawn area, there are several simple ways of feeding birds. Traditionally, household scraps – particularly crusts of bread were scattered on a lawn for birds to eat. This is both unsightly and can also easily attract rats. But food scattered on a lawn is important because many species prefer to feed in the open, where they feel safer from cats and other predators. The simplest solution to the rat issue, is, since rats are generally nocturnal, to ensure that only enough food for a day's bird feeding is put out.

A simple bird table, little more than a tray (with drainage) on a post, provides the simplest and most basic feeding station. This can be enhanced by suspending plastic nets of peanuts, coconut halves, and bacon rinds – all fairly straightforward. In fact, nearly all other feeding stations derive from this basic design. However elaborate they seem, most feeding stations, even the most expensive stainless steel and polycarbonate models, can be made at home, quite cheaply, and are based around just a few designs. But, since one of the main reasons for attracting birds is aesthetic, we generally like to have attractive-looking feeders. So the choice of design will reflect the owners' concepts of beauty, be it a rustic bird table, or ultra modern Perspex™ and chrome-plated, or quirky and home-made from recycled materials.

ABOVE: *A male Blackbird, which is not a bird normally seen on hanging feeders. This one has been ringed as part of a research project.*
LEFT: *A Crested Tit feeding on fat. In the British Isles the Crested Tit has a range restricted to the pine forests of Scotland, and is therefore a relatively rare bird at garden feeders. However, in many parts of continental Europe, it as a regular visitor to gardens.*

RIGHT: *A feeding station, with a wide range of different styles of feeders. Although there are many advantages to hanging feeders in trees, and many birds will find the setting attractive, there are also some disadvantages. Squirrels easily empty feeders hung in this way, and when the trees are covered with leaves birds may be vulnerable to predators such as cats. If you do hang feeders in this sort of situation, it is worth checking them at night, as they will probably also attract Wood Mice and Bank Voles.*

BELOW: *A lawn, particularly if it is not mown to a bowling green height, provides a good feeding area for several species. Pied Wagtails chase small insects, while Jackdaws probe for grubs, and Blackbirds hunt for worms. Doves feed mostly on seeds and vegetable matter.*

TOP: *Once fairly common in rural gardens in winter, the Reed Bunting seen here (**RIGHT**) with a female Brambling (**LEFT**) is now a comparative rarity. Like many other buntings and finches, Reed Buntings generally prefer feeding on the ground, in reasonably open spaces, where they have a good all round view for potential predators. A simple bird table, such as a large tree stump, is an ideal feeding station for such species.*

ABOVE: *One stage up from the ground or a tree stump, this feeding station is no more than easy-to-clean trays, just clear of the ground. The separate trays allow different foods to be supplied for the various species, and there is less wastage by birds scattering unwanted food while extracting preferred items. The Great Spotted Woodpecker is attracted to fat, the Great Tit to seeds.*

ABOVE: *An ideal feeding station. The flat platform is covered, so that scattered food does not become sodden in rain, and below a variety of different styles of feeders are hung, filled with differing foods that will attract as wide a range of birds as possible. The station is situated in the open, but with hedges and trees in the vicinity. This enables small birds to approach the station as they forage, but while feeding they have a clear view for potential predators.*

The type of foods and plantings to attract birds

The sorts of foods that attract birds are often – though by no means exclusively – similar to those they find in the wild. Most finches will be attracted by seeds, while insectivorous birds will be attracted by mealworms. However, some wild foods, such as caterpillars, grubs and other insects, are difficult to provide and substitutes need to be created. A common feature among many of the foods sought after by birds is a high energy content – peanuts and sunflower seeds are very good examples of this. In winter especially, when the days are short and the nights long and cold, birds need to gather as much energy as possible during the few hours available for feeding. And birds have a high metabolic rate. From a bird's point of view, the easier it is to obtain the food the better – shelled peanuts, peanut granules, shelled sunflower seeds, shredded coconuts and similar foods are all likely to be particularly attractive. However, most birds when presented with such foods, will grab as much as they can as quickly as they can and fly away to consume it in safety, away from the feeder. But the human observer wants the birds to stay in view as long as possible, so

ABOVE: *While Nuthatches will readily come for whole hazel nuts, they will then fly off to find a crevice in which to wedge the nut so that they can hammer out the kernel. In order to see Nuthatches clearly, put shelled nuts in a wire feeder to ensure they stay visible.*

ABOVE: *Redwings are mostly winter visitors and are especially fond of berries, travelling in flocks along hedges to find them.*

ABOVE: *Goldfinches are attracted by thistles, and will also come to feed on Nyjer, which is very similar and available from pet shops.*

ABOVE TOP: *Blue Tits feeding on Mealworms. Live foods are important for many birds, particularly during the breeding season, and are best fed from dishes with have covers, to ensure they do not flood and their contents are not drowned.* **ABOVE CENTRE:** *A Great Spotted Woodpecker feeding on fat. Bird cakes that contain fats, such as suet, are a particular favourite of woodpeckers.* **LEFT:** *By keeping foodstuffs separate and not feeding mixed seed, less is wasted. Blue and Great Tits (top feeder) prefer larger seeds and nuts, such as peanut granules, while Goldfinches (lower feeder) prefer small seeds, such as Nyjer.*

ABOVE: *The compost heap – or preferably collection of compost heaps in various stages of decomposition – is an essential part of any good garden, and is also invaluable for birds. A good compost heap will be full of worms and other invertebrates, all taking part in the process of decomposing the vegetation, and they will provide abundant food for insectivorous birds. A compost heap should also be turned occasionally, and at such times Robins will soon learn to come and feed.*

peanuts in their shells, or in a wire basket, coconut halves, sunflower seeds in their husks, or on the actual plant, are all ideal for keeping the birds in sight. There is a good reason why birds want to grab their food and leave – predators. It is therefore important to make birds feel as secure as possible while they are feeding.

Site feeders away from places where they will be constantly disturbed by passers by, and make sure the feeding birds have a good view around, and cannot be approached unawares by a neighbourhood cat, or a marauding Sparrowhawk.

Water is also important, particularly for drinking. Nearly all birds need access to fresh water usually several times a day. Additionally, many birds enjoy bathing. The importance of clean water cannot be over-emphasized, and is described in greater detail on page 64.

A garden is, in itself, a bird feeder. A healthy garden has its own ecosystem, albeit often a rather 'unnatural' one, dependent on constant management by its owners. Anyone who has gone away for more than a week, leaving a garden unattended in early summer, will know how rapidly a garden starts to revert to a more 'natural' state. As every gardener is aware, weeds

LEFT: *Wrens can be provided with foraging areas by allowing a dense cover of ivy to grow on walls, large trees and garden sheds. The dense evergreen foliage not only provides nesting sites for wrens and other birds, but is also usually full of insects and spiders, and the blossom, which appears in winter, attracts bees and other insects.*

larger the tree the better. We should also bear in mind that most gardens normally serve at least two or more functions. However keen a wildlife enthusiast you are, the garden is likely to be for aesthetic pleasure as well – it may also be somewhere for children to play, and to hold the occasional barbecue. But if you are planting a garden with birds in mind, then species such as hawthorns are ideal hedging,

always grow at least 10 times faster than garden plants. The soil of a garden will contain many species of invertebrates – from nematodes to earthworms, from woodlice to molluscs – and the vegetation of the garden will provide food and lodging for myriad insects and other invertebrates – some considered pests, others beneficial. But a healthy garden can achieve a balance and birds can be part of that equation. The grubs and caterpillars that might otherwise devastate a vegetable crop can be an important source of food for a pair of Great Tits rearing their brood of 10 babies. Piles of twigs left over from pruning, logs, and compost heaps are all full of insects, spiders and other invertebrates, which are good sources of natural wild foods for species such as Wrens and Dunnocks, which also find excellent hunting places in ivy-covered walls, areas of brambles, and overgrown patches of grasses.

Most books on wildlife gardening emphasize the importance of native species of plants. While it is certainly important to have as many of these as possible, I would discourage anyone from cutting down a large mature tree that is 'exotic' and replacing it with a small native species. Generally speaking the

providing a dazzling show of blossom in late spring, shelter, and a reliable crop of berries most years. However, exotic species such as firethorn, which have prolific and colourful berries are also useful. They are not so popular with most birds, and consequently last well into winter, providing food for late-arriving migrants – they are often associated with Waxwings. An important, and often overlooked, type of planting suitable for birds is the fruit and vegetable garden. Establishing lots of fruit trees (such as apples, plums and cherries), and also planting their wild relatives in hedges, will ensure high rates of fertilization and abundant crops. Share the fruit with the birds – in my experience it is actually very difficult to get the cherries before the birds do, but apples can be gathered and stored and any that start to rot can be put out in mid-winter for Fieldfares and other thrushes. Similarly, grow cultivated varieties of soft fruits, such as grapes, blackberries, raspberries and currants. You can put netting over some to ensure you get your portion, but do remember to share them with the birds. When vegetables, such as beets or parsnips, start to go to seed or 'bolt', leave some, as many birds will like the seeds. And you can grow many garden plants specifically for their seed. Sunflowers, thistles and teasels are just a few that make very attractive garden plants, and provide seeds for birds. TV gardener Monty Don is forever emphasizing the importance of working *with* plants, and in a garden designed for attracting birds and other wildlife this is paramount. By choosing plants that like the existing conditions, and not trying to grow sun-loving species in a shady garden or vice versa, nor heathers or other acid-loving plants on heavy clay and so forth, you will develop a more natural, balanced garden that will be easier to care for.

BELOW: *Tree Sparrows were once fairly common birds in outer suburban and rural areas, but have declined dramatically over the past 25 years. Like other sparrows and finches, they will feed on seeds and grain, generally preferring to feed on the ground.*

ABOVE: *Blackbirds are members of the thrush family, and like other thrushes will feed on rotting apples and other fruit in winter.*

Storage and hygiene

Seeds and grains should always be stored in conditions as dry as possible, preferably in airtight containers. While insect infestations may not be a serious problem – birds will usually eat the grubs — fungus and moulds can be more serious. Depending on the scale you are feeding suitable containers range from kitchen storage jars to dustbins with lockable lids (to stop them blowing off in a storm). For topping up the feeders a seed scoop is very useful – the types that combine the scoop with a funnel are the easiest to use.

It is important to keep feeders and bird baths clean. Each time they are refilled it is a good idea to give them a scrub. Many of the specialist bird feeding companies sell cleaning fluids, but an alternative is to buy anti-bacterial/viral cleaners from a poultry supplier. A bottle brush should be used to clean plastic tubes and any food that is decaying or rotting should be disposed of. Water for drinking and bathing should also be changed regularly. The husk and other detritus under the feeders is good fibre to add to a compost heap.

ABOVE: *Peanuts should always be purchased from a reliable source, and any that appear to be poor quality or going bad in the feeder should be discarded. Peanuts are prone to becoming contaminated with powerful, tasteless, odourless and colourless toxins, known as aflatoxins, which are caused by Aspergillus flavus fungi. Aflatoxins are carcinogenic and acutely toxic to birds, as well as humans, causing death, liver cancer, destruction of the immune system, reduced growth and reduced vigour. Since aflatoxins are absorbed into body fats and birds have low levels of these, birds are much more sensitive to aflatoxin than most other species.*

ABOVE: *Hygiene is important, and feeders should be cleaned regularly. Clear plastic tubes, such as the Droll Yankee brand are easy to clean, and most suppliers sell brushes suitable for cleaning the tubes.*

Should you feed in the breeding season?

For a long time many experts advised against feeding birds during the breeding season. This was because it was thought that birds would be feeding their babies on bread and peanuts and other unsuitable foods when they should be giving them grubs, caterpillars and other insects. More thorough research has shown that these concerns are generally unfounded, and that, even when birds do utilize such foods, they normally form only a very small part of the total diet. So it is generally thought safe to put food out all the year round. However, the consumption will normally fall dramatically during the summer months, for two main reasons: the first is that most birds visiting feeders change their diet, to take a much larger proportion of insects, and the second is that birds disperse more widely to establish breeding territories. More important than putting out food for birds during the breeding season is to make sure that natural food in the form of invertebrates is available. This means planting a wide range of shrubs, flowers and trees, and avoiding the use of pesticides.

ABOVE: *Great Tits feed their young almost exclusively on live food, particularly caterpillars and insect grubs. Like other tits, the young birds often stay together in small flocks when they have left the nest boxes or holes where they were reared. As soon as they are fledged they will start visiting bird feeders, where they can be distinguished from the adults by their yellower colouring.*

LEFT: *Robins feed their young almost exclusively on live food, and continue to do so for several days after the young have left the nest. The young robins are quite unlike the parents – being brown and heavily spotted. Looking like miniature baby Blackbirds, their rather thrush-like markings are an indication that they are members of the same family.*

Where to place feeders and food plants in the garden

Assuming you want to attract as wide a variety of birds as possible, an ideal garden for feeding birds is one that has as wide a variety of microhabitats as is possible – ideally open meadows, trees, hedges, flowering plants, cliffs, marshes and ponds. In smaller gardens it is obviously not possible to have all of these, but you can generally compensate by using artificial means. Birds are not too fussy about using man-made objects,

which is why some species have become successful as commensals (i.e. living alongside humans).

House Martins originally nested on cliffs, but found that houses, churches and other buildings provided similar vertical faces; just as Swifts originally nested in tree holes and cliffs, but found that the eaves of houses provided even better nest sites. When nestboxes were being made at the end of the

LEFT: *Even a balcony high above street level is suitable for feeding birds. House Sparrows seem to be disappearing from the city centres where they were once abundant, but tits are still found in all but the most heavily built up areas, and will regularly visit feeding stations, particularly if there are shrubs and other plants in containers.* **OPPOSITE:** *An idealized bird feeding garden. This garden assumes that the owners will want their own space, and close to the house is a paved patio area, on which to sit and relax, and dine. It is also an area where food can be put on the ground, but cleared away if it becomes stale or unsightly. Key features of the garden are large areas of unmown grass, but with mown paths for access, several large fruit trees, and plenty of shrubs, bushes and hedges. At the rear of the garden compost heaps and log piles provide important habitat, but perhaps the most important feature of all is as large a pond as possible. This basic design shows that it is perfectly feasible to create an attractive pleasure garden for the owner, full of interest and beautiful plants, but at the same time maximizing its value to birds.*

A) Long grass provides grasshoppers, beetles and other invertebrate foods
B) Compost heap provides worms, grubs and other invertebrates, as well as edible vegetation
C) Fruit trees provide apples and other fruits for winter feeding, and locations for nest boxes and other feeding stations
D) Hedges provide nuts and berries
E) Mown paths provide areas where Blackbirds and thrushes can forage for worms
F) Plants and shrubs provide foraging areas for Dunnocks, Robins and Wrens to feed on insects, spiders and other invertebrates
G) Pond provides permanent source of drinking water, and also a wide range of insects (including mosquitos)
H) Fountain provides movement in water to attract passing migrant birds
I) Feeding station and bird bath attract birds close to the residence, where they can be more easily observed
J) Terrace provides area for feeding ground-feeding birds, safe from predation

LEFT: *A wild flower border, using native species, can be an extremely attractive garden feature. Illustrated here are yarrow, cornflower, lavender, foxglove, crab apple, salvia and marigolds, most of which also provide seeds or fruits for birds.*

nineteenth century, they were fashioned out of hollowed logs and made to look as much as possible like a tree-trunk, but nowadays we rarely bother. Nestboxes, whether made of wood or of concrete or other man-made materials, bear little resemblance to natural cavities. The same applies to feeders. Rustic-looking bird tables may be slightly more aesthetically pleasing, but a plastic tray is probably more hygienic, and will feed the birds just as well. Feeders slung from brackets on a wall will attract as many birds as those hung from a tree. But there is no doubt that combining artificial feeders with natural sources of food, such as berry-rich hedges, and insect-rich ivy-clad walls is always the best option.

Which birds visit feeders?

In the UK the common birds at feeders and in gardens include Collared Dove, Woodpigeon, Feral Pigeon, Magpie, Blackbird, Starling, House Sparrow, Dunnock, Robin, Great Tit, Blue Tit, Coal Tit, Chaffinch and Greenfinch. These are the birds that can be expected in all but the most built-up surroundings, even though House Sparrows seem to be disappearing from many areas.

In suburban areas and rural gardens a lot more species are relatively common, and one can reasonably expect to attract many of the following species: Black-headed Gull, Pheasant, Carrion Crow, Jackdaw, Jay, Mistle Thrush, Sparrowhawk, Fieldfare, Redwing, Great Spotted Woodpecker, Nuthatch,

BELOW: *Magpies are increasingly common visitors at bird tables, yet prior to the middle of the 20th century, they were rarities. Although perceived as aggressive and destructive (they do kill small birds and take their eggs), evidence suggests they have little impact on the overall populations of small birds. This one is at a feeding hopper made from a drainpipe.*

Wren, Bullfinch, Goldfinch, Blackcap, Pied Wagtail, Marsh Tit and Long-tailed Tit.

Then there are many other species, which while not exactly rare, are not often seen in gardens or at feeders, or if they are, have a fairly localized distribution. These include Red Kite, Mallard, Moorhen, Lesser Black-backed Gull, Siskin, Redpoll, Yellowhammer, Goldcrest, Chiffchaff, Linnet, Tawny Owl, Treecreeper and Rook. And even such species as Water Rail, Red-legged Partridge, Mandarin Duck and Ring-necked Parakeets, all of which I have seen feeding in gardens.

There is another group of birds – those which visit gardens but rarely go to feeders and bird tables – such as Kestrel, Green Woodpecker, Cuckoo, Pied Flycatcher, Spotted Flycatcher, House Martin, Swift and Swallow. Finally, there are the real rarities, such as Waxwing, Hoopoe and Black Redstart, and a range of warblers, all of which can occur in gardens.

The species found in gardens are constantly changing. For example, before the 1960s there were no Collared Doves in British gardens – they had not

yet colonized widely, and until the 1980s Reed Buntings were not uncommon in rural gardens. And in the 1960s I can recall visiting gardens in southern England to see breeding Wrynecks – now quite extinct, and also finding Red-backed Shrikes nesting in suburban London.

While the above lists are by no means exhaustive, they do give an idea of the range and variety of the birds that can be attracted to a garden. Even those species that do not normally visit feeders will often be encouraged by the presence of other birds, and an added incentive for almost all birds is water.

ABOVE: *The Siskin is a small finch that was once a rarity in gardens, but over the past 50 years has become a regular visitor. Peanuts and small seeds are particularly good at attracting them, and they are identified by their rather yellowish, streaked plumage.*

ABOVE: *The Spotted Flycatcher is a species that appears to be declining rapidly in many parts of Britain. Although they are rarely seen at feeders, they often nest in boxes, or even in crevices in houses, and ensuring a plentiful supply of flying insects will benefit them.* **BELOW:** *Chaffinches will come to feeders and to bird tables, such as shown here, but given the choice they generally prefer to feed on seed scattered on the ground. They are one of the most widespread of all birds in the British Isles.*

ABOVE: *In the last 50 years Tree Sparrow numbers have plummeted. If Tree Sparrows occur in gardens, they should be encouraged, and food provided in such a way that more aggressive species will not discourage them.*

RIGHT: *Two Blue Tits and a Robin feeding on fat balls. These are among the easiest birds to attract to feeders and often the tamest. In fact, with patience, they can become tame enough to take food from the hand.*

Bird Behaviour

The daily routine

An important fact to remember about birds is that they are mostly diurnal – that is to say they are active during daylight hours: in summer, this activity starts very early. At all times of the year, birds are most active at dawn and also shortly before dusk. So if you really want to get the best out of your bird feeders, then making sure they are full and ready for dawn activity, and getting up soon after dawn, is worth the experience. While, in many parts of the country, the dawn chorus is not as intensive as it once was, it is still worth listening to – but this can mean rising around 4am or even earlier in the north of England or Scotland. As soon as they are awake, birds will start foraging, but in summer, particularly in hot weather, they will often rest in the shade during the middle of the day, becoming active again in the evening.

BELOW: *A simple bird table with an ample supply of seed and other foods shows that even the aggressive Greenfinches do not deter all other birds, such as the Nuthatch and Robin seen here. Provided there is a good supply of food, once satiated, the greediest of birds will move on and make way for other species.*

The annual cycle

A large proportion of the birds we see in our gardens are migratory, coming to Britain to breed and moving south for the winter – sometimes as far as Central or Southern Africa. Even birds that we think of as resident often move away in winter to be replaced by immigrants from Scandinavia and Russia: for example, the Starlings we see in huge numbers in winter are mostly from Russia. In the 1960s, I recall ringing a young Song Thrush in my suburban London garden, which was later found in Portugal.

We know that birds breed in spring, but spring can stretch over a long period. In a mild winter some birds start in February, while some migrants, such as Spotted Flycatchers, do not nest until late May. And several species have multiple broods. As soon as one brood has left the nest (fledged) another follows. Swallows, Blue Tits and Wrens all try to have several broods.

So for birds, spring and summer often merge into one extended breeding season, although most birds try to ensure that their babies are around when there is a good food supply. And most, including species such as sparrows, feed their young on insects and other live food. After the breeding season, the migratory species feed on high energy foods to build

RIGHT: *The type of beak of each species of bird gives an indication of the type of foods they prefer. The Mallard (bottom right) normally feeds by dabbling and filtering food in water, using a broad beak, whereas the Green Woodpecker (centre) uses its powerful beak for excavating grubs from decaying wood, or in anthills. The Wren (centre left) has a fine beak, which is ideal for gleaning insects and spiders in foliage, while the Blackbird (bottom left), has a larger, similarly proportioned beak, that is ideal for gathering insects and worms in leaf litter, as well as worms in lawns. The short, stubby beak of the Chaffinch (top left) is characteristic of finches, and used for cracking open seeds. The Jackdaw (top right) has a general purpose beak that can deal with carrion, as well as insects and fruits.*

up fat, in order to sustain them through their migrations. A wide range of species will form feeding flocks by autumn: tits often form family flocks, which might join together with other families and even mix with other species of tits. These flocks will visit bird feeders throughout the winter.

Feeding behaviour

Birds have a wide range of feeding behaviours, and it is one of the reasons they do not all compete for the same foods. A good guide to the feeding behaviour of most species is the bill – its shape and form often gives an indication as to how it feeds. The Mallard's broad shovel-shaped bill is ideal for dabbling on the surface of the water, and sieving food from the water. The powerful, pointed bill of the Green Woodpecker is used as a chisel for hammering trees to extract grubs, and also for probing anthills. The slender bill of the Wren is ideal for picking out spiders and insects in crevices, and hiding among dense foliage. Both the Jackdaw and the Blackbird have rather all-purpose beaks, though the Blackbird's, being small and less robust, is used for pulling up Earthworms and feeding on grubs, while the Jackdaw's is strong enough to be used for tearing at carrion, as well as being used as a probe.

Finally, two of the most common birds at the bird table have relatively small bills. The Blue Tit's short bill, which is ideal for feeding through mesh on feeders, is actually designed for picking insects from leaves and bark crevices, while the Chaffinch's bill, is sharp and pointed, but stout and ideal for getting seeds from their husks.

Territory and aggression

Many birds are territorial and not only during the breeding season. Towards the end of winter, even winter migrants will start behaving competitively, singing and fighting. Robins will fight all intruders, and some species, such as Pied Wagtails, become so aggressive that they will even attack their reflection in a parked car's hub cap. The most attractive feature of birds' territoriality is that this is the reason they sing – to proclaim their territory.

Some species are aggressive on feeders – keeping away other species – and within flocks there may be a 'pecking order', with certain individuals that are dominant always getting the best positions on the feeder, but this is usually difficult to observe unless the individuals can be recognised. Many people become annoyed when they see their feeders dominated by aggressive sparrows, Starlings or Greenfinches, but there are easy remedies: change the style of feeder and the type of food. Let the Greenfinches and the others have their favourite food, but put out more feeders, each with only one variety of food in it, and the less aggressive species such as Goldfinches will be able to feed in peace.

Nesting

While discussing bird feeding it is worth mentioning nesting behaviour, as providing nest boxes and other artificial nest sites is as good a way of attracting birds as feeding them. As mentioned elsewhere in this book, larger birds need larger territories and it is also difficult to provide for most birds of prey on the bird table. However, they too can be attracted with nest sites. The provision of nest boxes for wild birds goes way back in history, though in earlier times it was to ensure the birds were easily available for the pot. Baron von Berlepsch (p119), as well as being a pioneer in bird feeding, also created a range of nest boxes, the design of which is often remarkably similar to those still being made. Most nest boxes can easily made by an enthusiastic DIY-er, but there are some

ABOVE: *A Blue Tit entering its nest box. This is a well-constructed box with a metal exclusion plate. It is designed to prevent Great Spotted Woodpeckers enlarging the holes and eating their young.*

ABOVE: *A Robin at an open-fronted nest box. Although this is generally the preferred type of nest box for Robins, they are also known to utilize a wide range of cavities from old kettles to wellington boots hanging in a shed.*

excellent commercially-made types now on the market. While some of these are relatively expensive many of them have extremely long life expectancy, particularly those made from compounds based on concrete.

Nest boxes for small birds fall broadly into two categories: open-fronted, or hole nests. Robins, Flycatchers and Wagtails are among the species that prefer open-fronted boxes, while most tits use hole nest boxes. If you do buy nest boxes it is important to buy a good one. Those sold for aviary birds are often quite unsuitable for wild birds, and those sold in supermarkets are also often useless. A nest box needs to be thick enough to maintain a steady temperature – if it is too thin, the temperature can become too hot, and the babies will die from heat exhaustion. The hole not only needs to be the right size, but also to be high enough up to prevent predators, such as magpies, yanking out the chicks. It is always best to buy nest boxes and bird foods from specialist wild bird suppliers, or the nature centres of the RSPB and similar organisations.

Nesting material

Do not be tempted to stuff a nest box with nesting material – it will probably just put off any potential occupants. Each species has their own very distinctive design –even though it may appear fairly random and sometimes include items discarded by humans. The best way of helping nesting birds is to ensure there is a supply of the sorts of materials they use and let them help themselves and build their own nests. Short pieces of cotton or wool, cut hair, combings from dog grooming, hay and straw are all very useful. You could even keep thistle down from one year to the next, and the contents of an old feather pillow will be eagerly sought after. Nesting material can either be scattered, which is usually rather unsightly, or placed in wire feeders for the birds to extract.

NESTING MATERIALS

During the breeding season, in addition to food you can help nesting birds by putting out nest building materials. House Martins need a ready supply of good quality mud (i.e. with plenty of clay) around the margin of a pond. The hair from grooming a pet dog is soon taken by a wide range of birds for lining the nest, but straw, short pieces of string, grass clippings, short pieces of cotton (short, so that birds do not become entangled) and feathers from old pillows will all be welcomed. These can be placed in a mesh feeder, and the birds will extract the bits they need.

ABOVE: *This type of artificial nest is commonly sold in pet shops, and is designed for caged birds. They are not really suitable for wild birds and are generally best avoided.*

How to attract birds

One of the best ways of attracting specific birds to a garden is by looking at the food requirements of the birds listed on p29, and seeing how best they can be fulfilled. Any good handbook to British birds will give their food preferences, and the mammoth nine-volume Birds of the Western Palearctic gives it all in extreme detail.

But feeding birds should not be simply about attracting large numbers of birds. There may be pleasure to be gained from watching large flocks of Starlings all the time, or a constant stream of Blue and Great Tits, but arguably not as much pleasure as seeing 20 or more different species. To maximize the diversity of species, the habitat as well as the foodstuffs offered must be as varied as possible, and consequently the layout and design of a garden is an even more important aspect of bird feeding. The bird gardener is lucky if they own a garden with well established mature trees, particularly if they are fruit

RIGHT: *Bullfinch numbers have declined by over 50% in recent years, but they were once extensively persecuted because of the damage they did to the flower buds of commercial fruit crops. Although they do eat fruit tree buds, in most gardens it has little or no impact on the final crop.* **BELOW:** *In autumn, Jays become very mobile; oak trees are likely to attract them, as acorns are a favourite food. For the rest of the year they are usually very secretive, preferring larger parks and woodlands.*

or nut trees. There are few trees better for attracting a wide range of birds than a large mature oak, old apple tree, large hazel coppice – but almost any large tree has wildlife value. Tits, warblers and other small birds will glean among the foliage, while Jays, woodpeckers and Nuthatches will gather at other times of the year. Even a bare open plot can, with thoughtful planting, soon become a haven for birds and other wildlife. The Directory (p84) has examples of many of the best trees and shrubs but remember to consult a good gardening manual, ensure you have the right soil types, and the right amount of sunlight and moisture for the species you want to plant. A sun-loving, acid-soil species is unlikely to flourish in a shady corner of a chalky hill.

Choose a few fast-growing species too; even the infamous *leylandii* cypress can have a place as a temporary protection in an open garden, providing cover while other species grow in its shelter. It also provides good habitat for spiders and other small invertebrates for crests and warblers to feed on. For the bird gardener, fruit trees are a must. There's enjoyment for yourself, your friends and neighbours, and plenty for the birds. By planting a selection of apples, pears, medlars, cherries, mulberries and other fruiting trees, according to the size of the garden, you are guaranteed a beautiful show of blossom in spring, followed by a fruit crop for you and the birds in the autumn. You should not have to wait too long, as most fruit trees are sold at a size when they will start fruiting almost immediately.

Finally, if you are planting lots of trees in a large garden, plant them small (as one- or two-year-old 'whips') and plant them densely. Smaller trees put down healthier roots and within a few years will have caught up with the larger more expensive pot-grown trees. It is also easier to thin trees than fill gaps later.

There are a few key elements to bear in mind when assessing the value of a garden to birds. Remember, birds generally do not like large expanses of closely mown lawns. While there are exceptions to this, the overwhelming majority prefer grass to be left to more than bowling-green height, and also prefer it to be interspersed with shrubs and herbage. Remember to look at neighbours' gardens – there is no point in duplicating features for the sake of it, if space is limited. Try to look at the habitat as a whole, just as a bird will.

BELOW: *In larger gardens with orchards, it is a good idea to leave some of the fruit crop for birds to feed on in winter. Fieldfares and other thrushes (including Blackbirds) all feed on the rotting fruits at a time of the year when there is often little else to eat.*

Keeping undesirables at bay – rats, squirrels and aggressive birds

Feeding birds will attract a wide range of other wildlife – even muntjac deer have been seen standing on their hind legs to feed at a bird table. The problem is that we humans are discriminatory in the species we wish to attract. Very few people want to see Brown Rats on their bird table; however intelligent and amusing they are, they are still serious pests and disease carriers. For some reason that I personally find difficult to understand many people seem to have an almost pathological hatred of Grey Squirrels. Whole books have been written about keeping them off bird tables

and out of gardens. I like squirrels, and think it should be possible to come to some sort of arrangement for making some food available to squirrels, but some off limits, as they can be very greedy and do tend to drive some birds away.

More controversial are predatory birds. The RSPB has had to confront this problem frequently when even its members demand that 'something should be done about Sparrowhawks'. From near extinction in the 1950s and 1960s due to pesticide poisoning, Sparrowhawk numbers have now recovered significantly, at the same time as the once incredibly abundant House Sparrow has plummeted in numbers. Sparrowhawks are often quick to learn that bird tables are good sources of food. The reality is that predators rarely have significant impacts on their prey populations, but it is difficult to believe this when day after day a Sparrowhawk swoops though your garden grabbing a Blue Tit as it passes. Ensuring that feeders are close to dense shrubs, particularly those that are dense and thorny, into which small birds can retreat, is a good strategy.

Sparrowhawks rely largely on the element of surprise, dashing through trees and bushes, and swinging round buildings. If you find that Sparrowhawks are becoming a bit too enthusiastic, simply moving the bird feeders may be all that is needed to thwart their success. Reflective globes on a pole are also thought to be effective deterrents, as are dangling CDs (use the free samples) although rather unsightly.

Many other birds can become pests, simply because they are large, aggressive and greedy, but relatively few actually attack other birds at feeding stations. In some cities gulls can be a problem and, occasionally, because of their sheer numbers and appetite, so can Starlings. All one can do in such circumstances is to stop supplying whatever it is that they are eating, and concentrate on food for smaller

LEFT: *The introduced Grey Squirrel (originally from North America) is very much part of the fauna. Though loved by many, it can be a pest at a bird feeder, simply because it is greedy, and loves to carry off food and hide it.*

RIGHT: *Magpies do feed on small birds, but they do not appear to have long-term effects on the populations, however distressing it is to see a young bird being dragged from its nest. The increase in Magpies has often been attributed to the increase in food available in the form of road casualties.*

ABOVE: *There are various feeders designed to keep out larger birds such as Starlings, but even these do not always deter Sparrowhawks, which can sometimes snatch a small bird through the wires.*

birds, in feeders designed to be difficult for larger ones to use. Other species that cause complaints include Greenfinches, which can be aggressive towards other birds at a feeder. Yet some species that can be aggressive, such as Great Spotted Woodpeckers, are generally tolerated, presumably because people like seeing them and they are only present in small numbers. When confronted with interspecific aggression in birds, there are two ways of resolving the problems – one is to put up more feeders, in a wider range of situations, and the second is to grow to like the transgressors. After all, Starlings, with their iridescent summer plumage, and

heavily spotted winter plumage, are actually very attractive birds. And Greenfinches are surprisingly diverse, with males and females moulting their way through several variations of plumage until fully adult. By erecting specialist feeders, such as thistle feeders, near but not too close to sunflower feeders, you should ensure that the Goldfinches can feed in peace while the Greenfinches gorge themselves on sunflower seeds.

My personal 'pest' species of bird, which I would discourage from feeders, is the Pheasant. In some parts of the British countryside it is bred in huge numbers and released into the wild for shooting, and because of its carefully protected status outside the hunting season, is often very abundant. It is a major pest to wildlife, eating almost anything small that moves, including lizards and Slow-worms, which are becoming increasingly rare.

There will always be some birds and other wildlife that people will want to keep away from their feeders. One of the simplest ways of excluding larger birds is to enclose the entire bird table or feeding station in a wire mesh cage, with a mesh size of 5 cm (2 in), which will allow tits and other small birds to pass through it easily but exclude larger species, such as Starlings, crows and pigeons. It will also help prevent attacks by Sparrowhawks. Herons, particularly young ones, often find that a garden pond provides

ABOVE: *While most people would like the idea of a heron visiting their garden, the loss of their pet goldfish is not so attractive. Fortunately, it is relatively easy to prevent predation, particularly in small gardens with small garden ponds. Larger ponds and lakes can normally sustain predation by herons.*

ABOVE: *Male Pheasants are spectacularly colourful, and the first few times they visit a garden, provide excitement. However, they can be greedy feeders and they are destructive to other wildlife, such as lizards, Slow-worms and amphibians. For this reason they should be discouraged.*

comparatively easy pickings and may therefore be unwelcome visitors if you keep ornamental fish. Herons tend to be very shy and are most likely to raid garden ponds at dawn and dusk. They prefer ponds with shallow, gently sloping banks, so if goldfish and koi carp are essential elements to your garden, they are less likely to be snatched by a heron if the pond has sheer, steep sides. Wire or string barriers about 20 and 35 cm high can also be placed around the edge of the pond. There are a number of other deterrents you can try from netting the pond, and giving the fish plenty of cover, to putting up scarecrows or even commercial bird-scarers activated by trip-wires. Plastic herons are not a successful deterrent, because herons often feed in groups.

Rats and squirrels can be prevented from climbing up the supporting post of a bird table by a smooth conical shield (*opposite*). But it is often astounding how agile they can be, particularly squirrels. They will negotiate thin wires, gnaw through strings to carry off entire feeders, and show what appears to be significant intelligence. So why not enjoy their antics? Rats are another matter and, while the odd one may be acceptable, provided you live in a fairly rural area, should there be any signs of numbers increasing at all, then it is best to contact the local pest control officer and have them dealt with humanely and professionally.

ABOVE: *Despite their ingenuity, squirrels can be prevented from getting access to a bird table by using a baffle, such as the one illustrated. These are sold by many of the suppliers of bird feeders, or can be home made; they also keep Brown Rats at bay.* **RIGHT:** *This type of feeder – shown with a Blue Tit (left) and a Great Tit (right) – deters many of the larger birds, such as Starlings, by making access to the food difficult, so that only the smaller and more agile tits can feed.*

Hazards in the garden

There are three main hazards to birds in a garden that have been introduced by man and should be borne in mind when siting feeders. They are cats, windows and traffic.

I, along with many other birders, have a reputation for being an ailurophobe (cat-hater). This is not true, but I do not believe that domestic cats should be allowed to roam freely where there is any risk of them killing wildlife. They should be kept under control, just as we expect dogs to be under control. It is often said that the best way of protecting a garden from invasion by local cats is to own a dog, and while this

RIGHT: *Cats are significant predators of garden wildlife, including birds, but there are ways of reducing their impact – even the choice of breed and colour can affect the ability to kill. Fluffy Persian types are generally less of a threat than tabby, wild types.*

BELOW: *It is important to ensure that cats cannot approach feeding areas without being seen. Most birds are able to escape if they have sufficient warning. But when birds are feeding newly fledged young, they are most vulnerable, and cats should not be roaming free.*

ABOVE: *A Song Thrush brooding. It is very important to leave pruning and even trimming of hedges and trees until the breeding season has finished, and remember, many birds raise two or even three broods, and they may be nesting until late summer.*

is probably true, it is not always practical. However, there are numerous humane sprays and electrical devices now being marketed for deterring unwanted cats, and the RSPB even has a section of its website devoted to this admirable cause.

Reflections of the sky and garden in windows seem to be one of the reasons that birds fly into them. A straight line of sight through a pair of windows can be particularly hazardous. Curtains can help to reduce these dangers, and silhouettes of hawks can be used with varying success.

Perimeter hedges can also be a source of danger to birds, particularly if they border well used roads, where birds may fly from hedge to hedge, not realizing that they are in danger of being struck by passing cars. And if you are using netting to protect soft fruits or vegetables it is important to use a fairly heavy gauge to reduce the likelihood of birds becoming entangled. It may be stating the obvious but try to avoid disturbance to breeding birds, such as by leaving the trimming of hedges and the pruning of shrubs until after the nesting season.

Finally, there is light pollution. I recently arrived in the middle of Norwich in the middle of the night, and four hours before dawn heard Blackbird, Robin and Dunnock all in full song. Flying over Europe a few hours earlier, the level of light pollution had been astounding and it made me wonder how it affected migrating birds; it was certainly affecting breeding birds. There is no doubt that light pollution is having negative impacts on a range of wildlife, so do your bit – switch off unnecessary exterior lights and save energy as well.

Getting serious about bird feeding

Anyone who is interested in birds really should belong to the Royal Society for the Protection of Birds (RSPB). Its magazine is packed with useful information together with advertisements for most of the leading suppliers of bird foods and other related items, such as binoculars. The RSPB also has numerous local groups, which will welcome anyone, whatever their level of expertise.

For the even more seriously interested there is the British Trust for Ornithology, which organizes scientific studies, including the national ringing schemes for birds. Additionally nearly all counties and many larger towns have bird clubs and natural history societies – all keen to increase their membership. The county bird clubs and natural history societies usually produce annual bird reports, summarizing the observations of their members. You can usually find out about these from the internet or your local public library.

Binoculars

If you are serious about bird feeding you will want to know what birds are coming to your garden. While most species can be easily identified, even without

ABOVE: *The great thing about bird feeding is that the birds come to you, so relaxing on a deck, for afternoon tea, binoculars in hand, is the best way of enjoying your bird feeders.*

ABOVE: *Whitethroats are one of the more unusual birds that can be attracted to gardens by planting the right bushes. Best of all is a large bramble patch: this provides a good nesting place and then an abundant supply of fruit, which Whitethroats are particularly attracted to, in late summer.*

binoculars, some of the more unusual and shyer species are more easily identified with the aid of binoculars. Choosing binoculars can be a bit of a minefield; there are so many, at such a wide range of prices, to choose from. But the first rule to remember is that more powerful is not necessarily a good thing, particularly for garden birdwatching. A pair of binoculars with 12× magnification may seem to be a better bet than 6× or 8×, but they will be much more difficult to hold steady, and do not focus as close, which can be a major disadvantage in a garden. A pair of 8×30 binoculars means a magnification of 8 times, and an objective lens of 30 mm. The larger the objective lens, the greater the light-gathering capacity, and theoretically the clearer the image. Quality is important, but generally it is recommended that a beginner should start with binoculars of 8×30 or 8×40 or possibly a magnification of 7×. It is best to avoid cheap binoculars, particularly those with high magnifications. Most serious birdwatchers move to a

LEFT: *Although once incredibly abundant, the House Sparrow has been declining in Britain since the early part of the 20th century. With the disappearance of horses from cities in the 1950s, its decline continued, until it actually started to disappear completely from some areas in the last years of the 20th century.*

BELOW: *Tree Sparrows are similar to the male House Sparrow (above), but the sexes are similar and quite distinct from the female House Sparrow (left). Although never as common as the House Sparrow, it is now very rare even in rural areas.*

telescope on a tripod if they want high magnification. Really expensive binoculars are often not significantly better than the mid-priced varieties, but if you visit one of the larger nature reserves of the RSPB they can often advise, and you can see for yourself in their shops.

Identification guides

The first of the modern-style field guides was published in 1954, and some birders still use the revised editions of this 'Peterson' guide. However, there is now a plethora of excellent guides available. Like binoculars, big is not always best. Many of the best field guides are designed for experienced birders, and cover a large area – the whole of Europe or the Western Palearctic (which includes North Africa and the Middle East). For the inexperienced garden

birdwatcher, this simply introduces dozens of species, never likely to be seen in a garden on a feeder. So choose a guide that only deals with British birds. Most nature reserves will sell simple identification charts for the most common species, which you can pin up by the window you use most often for watching the birds.

Garden bird feeding survey

The British Trust for Ornithology (BTO) organizes the longest running annual survey of garden birds in Europe (it was launched in the winter of 1970/71). Around 250 birdwatchers make weekly counts of birds coming to feeders between October and March. Its most important contribution has been that the data gathered showed that apparently common species such as Starling and House Sparrow were in serious decline. This was only detectable because of the long series of counts from a range of habitats. For example, the average maximum number of Starlings counted at bird feeders in suburban gardens fell from 16 in 1970 to 6 in 2002, but because of fluctuations, this is really only apparent when looked at over a period of time.

ABOVE: *Chaffinches are among the most common and most widespread birds, being found in over 90% of gardens. The long-term monitoring of their numbers seen at bird feeders provides vital data for assessing trends in the health of wild bird populations.*

LEFT: *Blackbirds are ubiquitous garden birds, and found in almost every garden surveyed. However, their territory may extend over several gardens, as they need quite a large area in order to forage for worms and other invertebrates.*

RIGHT: *Blackcap numbers fluctuate from year to year, depending on the weather conditions, both in Britain and also in their wintering areas in Africa. Increasing numbers are spending the winter in southern England – possibly an effect of global warming.*

According to the results of the 2002 Garden Bird Survey, Robins, Blackbirds, Blue Tits and Great Tits are found in almost all gardens. The next most common species are Collared Dove, Chaffinch, Greenfinch and Dunnock, all of which are found in over 90% of gardens. Coal Tits, House Sparrows, Starlings and Magpies were recorded in between 70%–90% of gardens. These figures show some marked changes from the first survey in the 1970s. At that time the House Sparrow was found in 97% of gardens, but this percentage fell to only 87% by 2002; the Magpie, which was previously recorded in only 27% of gardens, had nearly trebled to 73%; and the Collared Dove, recorded in 60% of gardens in the 1970s was found in 92% by 2002.

The BTO Garden BirdWatch

In addition to the Garden Bird Feeding Survey, since 1995 the BTO has organized a much broader-based survey, which by 2003 had over 16,000 participants and was still growing. The idea behind the survey is to gather data that helps to monitor the changes in the use that birds make of gardens. Because of the size of the survey, much of the information has to be gathered in a way that it can be analysed electronically, and an increasing proportion of the data sheets are now submitted over the internet – the BTO website is well worth a visit for further information. Other features of the Garden BirdWatch website include a postcode search which allows the visitor to get a list of all the birds recorded in gardens in their local area.

Section 2

Bird Feeders and Feeding Stations

Bird feeders come in a wide range of forms – the simplest way of feeding birds is to throw the food on a lawn, patio or window sill. But this has several disadvantages, not least the tendency to attract unwanted scavengers such as rats, and it can be messy and unsightly. The obvious solution to this is a bird table, which in its simplest form is just a tray, with drainage holes, on a post. Food can be put on the tray, and also hung from it.

ABOVE: *Seeds provide high energy sources of food for birds, which supplement their natural diet, and are particularly important in winter.* **OPPOSITE:** *Great and Blue Tits on a stainless steel feeder, which contains the nuts and makes them difficult to carry off, so the birds remain in view while feeding.*

Bird tables

Bird tables can be elevated on a stand or a feeding platform on the ground – many birds, such as thrushes, Starlings and Wrens, prefer feeding close to the ground, while others, particularly tits, prefer to be higher up. Bird tables can also be hung from a tree or bracket, and although these are less stable, they may attract a different range of species. One of the main reasons for erecting a bird table is to make the birds being fed easier to see, so choose your site with care. Make sure it is not only clearly visible from the window you are most likely to want to watch from, but also in a predator-proof site. You don't want it too close to a bush where a cat can leap out on unsuspecting birds. Nor do you want it so far in the open that a Sparrowhawk can learn to surprise the feeding birds – once they have found a good ambush point, they will probably return time and again. A final hazard to be aware of is glass windows. Every year hundreds of thousands of birds are killed flying into windows (*see* page 43 for tips on minimizing the impact).

Ground-level tables

For species such as Robin and Dunnock, it is often best to put foods such as seeds actually on the ground, but it is probably a good idea to use some sort of dish to contain other types of food, such as minced meat or mixes for insect-eating birds. A dish

ABOVE: *Although food hoppers on the ground will attract many birds, such as Robins, the disadvantage of placing them on the ground is that they will probably attract rats and mice.*
BELOW: *A Blackbird, Great Tit and Blue Tit feeding from a tray. Raised from the ground, with plenty of drainage holes (ideally a perforated metal base) this is a simple option for attracting a wide range of birds (as well as mice and other wildlife).*

ABOVE: *A feeding tray, with a mesh base so that the seed is kept dry, containing a mixture of wheat, maize and sunflower kernels. A tray is the best way of feeding buntings and finches, as they prefer to be on the ground.*

can be cleaned regularly to help keep the risk of disease at bay and it will prevent live foods, such as mealworms, from immediately escaping by burrowing underground. Platforms, often made of perforated metal, are good for small seeds, and can be moved around a lawn. The perforations are for drainage and prevent the food from becoming waterlogged. Most ground level tables are just slightly elevated, which may also afford feeding birds some protection from predatory cats by giving them a better view of the surrounding ground.

Elevated tables

A post table, one supported against a wall on brackets, or one suspended from a bracket or branch all fulfil the same function – a platform on which to place food or from which to hang feeders. In general, feeding tubes are more satisfactory than most other types of feeders and tables, simply because they keep the food dry and fresh. However, they are not suitable for all types of food, and not all bird species will use them. Household scraps, such as stale bread and cheese rinds are always best placed on tables. Because of the British climate, a table generally benefits from a roof, although this may deter some of the shyer birds – and some sort of provision for drainage remains essential.

A good bird table is solid, firmly anchored to the ground, and has the feeding area covered to keep food dry. A large, well-built table can have food hung in baskets and feeding tubes fixed to it. Most garden centres sell bird tables, and if you are reasonably competent at DIY it is simple to make one. One of the most important factors when erecting a bird table is to ensure the upright post is fairly substantial – 10 cm (4 in) square or more, and if it is being buried in the ground, make sure it goes in nearly one third of the total length. Treat the wood with a preservative that is not harmful to wildlife, such as a creosote replacement. Metal and plastic bird tables are also possibilities – they have the advantage of being easy to keep clean, and the shiny metal poles are more difficult for rats and squirrels to climb. Metal collars are often used to keep squirrels off tables. One design that should always be avoided is a combined bird table and nestbox – definitely not a good idea.

ABOVE: *Blue Tits at a food hopper. Wooden hoppers like this one have been used to feed wild birds for over 120 years. They are easy to make, and easy to use, keeping a supply of food clean and dry.*

Finally, space is probably going to dictate the type of bird table you put up. If space is at a premium, you will almost certainly want to combine the table with some drinking water, as well as hanging feeders, but if you have plenty of room, site a variety of feeding and drinking stations well apart from each other so that the shyer birds will not be threatened by the bolder species.

LEFT: *Great Tits on a feeding table. The roof helps keep food dry, but many species of birds feel hemmed in and prefer open tables where they can see the sky and any predators.* **BELOW:** *A Siskin on a feeding table. Siskins are seed eaters, but as well as liking a range of small seeds, they also like peanuts, particularly peanut crumbs.*

Bird feeders

Garden centres and the visitor centres at most nature reserves sell a wide assortment of feeders, ranging in price from £1.00 to over £50. There are two main types of feeder – mesh and tubes. The mesh sort are designed to hold nuts, and the tubes are mostly for seeds. In addition, many specialist feeders have been created to hold slices of bread, pieces of fruit, balls of fat and other foods. While the mesh feeders are easy DIY projects, the tubes are less easy, though they can be made out of hollowed wood, pieces of drainpipe, or similar materials. It is also possible to buy metal 'trees' to hang feeders from – but if you have the space, spread the feeders around the garden.

ABOVE: *A feeding bell containing a fat ball attracts tits and other agile species, and the bell keeps the food dry in the rain. A cheaper version can easily be home made using a flowerpot.*

ABOVE: *Great Spotted Woodpeckers are particularly fond of suet and bird-cakes made with fats. This feeder combines three types of food, and is ideal in a small space, but generally it is best to separate feeders so that aggressive birds do not drive away the shyer species.*

RIGHT: *A Blue Tit on a predator-proof feeder. The large mesh prevents predators such as Sparrowhawks being able to catch small birds while they are feeding. It also helps reduce competition from larger, greedier birds such as Jays, pigeons, Starlings and Pheasants.*

Mesh food holders

Food holders made from galvanized wire mesh come in a variety of shapes and sizes, and are relatively easy to make by anyone with the slenderest of DIY abilities. You can make a holder for peanuts from a tube of fine wire mesh, closed at both ends with a wooden disc, or you can form the wire into a cube. A flat rectangle of wire can be used to hold a slice of bread. The idea is that the birds can feed on the contents without being able to fly off with whole nuts or chunks of bread. This type of design is also available in a wide range of commercial designs, which usually have the advantage that the ends are often made of plastic, and consequently are easy to keep clean and hygienic.

The birds that use these feeders include most species of tits, as well as Greenfinches, Great Spotted Woodpeckers and Nuthatches. Grey Squirrels are often clever enough to find out how to open them, and abscond with as much as they can carry off, so if squirrels are a problem, it is best to use a stainless steel model and make sure the top is securely wired shut.

Among the most popular feeders are clear plastic (polycarbonate) tubes, which come in a wide range of sizes – up to a metre or more in length, with a huge capacity. Clear tubes have several advantages: they are easily kept clean and disease free, and it is easy to see when they need refilling. The best ones have a tray at the bottom, since many birds are wasteful feeders, spilling quantities of seed. Tubes with seed

ABOVE: *Long-tailed Tits feeding on peanuts in a mesh container. Outside the breeding season Long-tailed Tits normally travel around in family parties, so it is not uncommon to see 10 or more together on a feeder. A feeder like this one is easily made from some wire mesh and lids from a jam-jar.*

ABOVE: *A Nuthatch feeding on its favourite food – nuts. Feeders of this type are available in shops in a range of designs. The outer wires help to prevent larger birds getting at the food that is contained within the mesh.*

RIGHT: *Although Robins are mostly insectivorous, feeding on insects and other small invertebrates, they will also take small seeds, such as millet.*

BELOW: *A Greenfinch and House Sparrow feeding on peanuts. This type of stainless steel feeder is ideal. It is robust, easy to clean, easy to fill, and will attract a wide range of finches, buntings and tits, as well as woodpeckers and Nuthatches.*

trays are particularly important if you are feeding birds with mixed seeds – each species will have its own preferences, and the tray catches discarded seeds for other birds to eat. But as described elsewhere, it is always preferable not to fill them with mixed seeds. It is best to have as many different designs of feeder as possible, as each species has its own preferred methods of feeding, and preferred foods. For example, finches will find it easier to use feeders that have perches, whereas over-zealous birds, such as Starlings, can be kept at bay by a range of deterrents.

ABOVE: *A large feeding station, such as that illustrated here, has the advantage that it is easy to maintain, because all the feeders are close together, but it may deter shyer birds. It is also important to ensure that it does not attract rats. While rats can be entertaining visitors in the short term, because of their fecundity and destructiveness, they are not likely to remain welcome visitors.*

LEFT: *DIY feeders are easy to make from plastic bottles. All you need is a pair of heavy duty scissors and a bit of imagination. And if they don't work, try another design. Adapters for plastic bottles are also available in shops.*

Window feeders

Plastic trays that can be fixed to a window with suction pads bring wild birds almost into the house, but care is needed to ensure they are not sited in a way that birds may fly into the window and injure or kill themselves.

Chicken feeders

Most birdwatchers and other people who feed their garden birds do not seem to be aware of the wide range of feeders available for cage and aviary birds. The smaller ones used for bantams are ideal for feeding grain and seed; they are relatively cheap (often a lot less expensive than those sold for wild birds) and some of the hoppers are designed to be suspended. Visit a pet supermarket or an agricultural supplier to check them out. This could be particularly useful if you have a larger garden and are feeding some of the larger bird species. You can also make considerable savings by buying feed such as wheat in bulk, by the 25-kg sack. The main advantage of chicken feeders is that they do not need refilling very often – they can be filled with enough seed or grain to last a week or two.

Another form of cheap feeder can be made from plastic bottles. Adapters, sold by bird food suppliers for little over £1.00, fit the screw top of a 1.5 or 2-litre plastic bottle, and turn it into a very serviceable feeder.

Plastic tubes

Although many tube feeders are now made of clear plastic, this is by no means essential – it just makes it easier to see when they need refilling. If you are creating a DIY version, plastic drain pipes and other plumbing pipes are perfectly suitable. A visit to a hardware store with a good stock of plumbing and drainage accessories should soon give inspiration to anyone with some basic DIY skills. Plastic pipes, joints, curves and covers, all lend themselves to being

ABOVE: *It should always be remembered that aesthetics are a part of bird feeding. There are so many different designs available that you should be able to indulge personal preferences and taste.*

LEFT, BOTTOM LEFT AND BOTTOM RIGHT:
Variations on a theme. These feeders all contain similar elements – they are all clear plastic, which makes it easy to see when they need refilling, and they all ensure a clear view of the birds when they are feeding. However, if you are using a selection of feeders, it is probably best not to fill them all with mixed seeds, as shown here, but to put a single variety of seed in each feeder, as this will prevent wastage. When eating mixed seeds, birds will often simply discard their less preferred varieties in order to get at the seeds they like.

adapted as bird feeders – often at a fraction of the cost of ready-made ones. Plastic pipes usually come in three different sizes – the largest for underground drainage, and the smallest for interior plumbing, with roof drainage and guttering in intermediate sizes. It is easy enough to make a very robust feeder from the following pieces of roof drainage pipes – two down-pipe joints, four guttering ends, a balloon (normally used to keep leaves from blocking a down-pipe), and a clip to fix to a wall or post. It can be assembled using any suitable glue, and once fitted together filled with a wide range of foods such as grain or mixed seeds. Smaller feeders can be made using interior plumbing accessories and a large one, suitable for doves, Moorhens, crows and other larger birds can be made from the underground drainage accessories. And if you wanted to, you could always find earthenware pipes, which look a lot more attractive than plastic. Although they are not so easy to keep clean, and are more easily broken, earthenware pipes could easily be built into garden features.

Ready-to-hang feeders

Peanuts are often sold in disposable red-mesh plastic nets or plastic mesh tubes, and there are a number of other foods that can be hung out on string or fixed to a tree or fence.

ABOVE: *Siskins feeding on peanuts in a red plastic net. For many years birdwatchers thought this was the best way of attracting Siskins as they seemed to have learned to associate the red plastic with food.*

ABOVE: *A female Great Spotted Woodpecker on a coconut. She lacks any red on the head, which distinguishes her from the male, which has a red patch on the nape.*

ABOVE: *Bird cake made of seeds attracts Great Tits and many other small birds. Such a cake can be made by filling a flowerpot with seeds and pouring melted fat over it to hold it together.*

*Suet (see p96) is an important high-energy food for birds in winter. It is very versatile in the way it can be fed . The log with holes drilled into it, (**ABOVE LEFT**) is a very old method first used by Baron von Berlepsch (see p119) over 100 years ago, and the hanging basket (**ABOVE CENTRE**) is ideal for slices of suet or bread, or bread soaked in fat. The fat-filled discs (**ABOVE RIGHT**), are more decorative than functional – but might make Christmas tree decoration before being given to the birds.*

RIGHT: *Coal Tits, like all other titmice, are particularly fond of fats, and the feeder cones and fat balls sold in supermarkets and pet suppliers are ideal for them.*

Red plastic nets filled with shelled peanuts once achieved a degree of fame as bird feeders, when it was noticed that Siskins were often attracted to them. The reason for this is not very clear, although it has been suggested that they were not dissimilar to their main natural food – alder cones. However, I find this association difficult to make.

Bird 'cakes' (*see p96*), fat balls, sprays of millet and seed 'bells' (sold in almost all pet shops), sunflower seed heads and coconut halves can all be hung out as ready-made feeders.

Suet is a high-energy food, appreciated by a wide range of birds, particularly in winter. It can be presented in many different ways, and very often the feeders are designed to prevent the larger, greedier birds taking it all and depriving the smaller species of their share, as well as preventing the viewer from seeing the birds. If you have a large tree, with creviced bark, such as Oak, the suet can be simply smeared in the cracks. On a smoother tree, such as Birch (**BOTTOM RIGHT**), a mesh holder can be used, easily made from an small off-cut or wire-netting.

Alternatively, a simple hanging feeder, suitable for tits and other small birds, can be made by melting the fat and pouring it into a large pine cone (**TOP RIGHT**). The Blue Tit (**TOP LEFT**) is feeding on a fat ball, of the type widely sold in supermarkets, pet shops and elsewhere. While it is possible to make them, they are so cheap to buy, it is hardly worthwhile. By using a little ingenuity and imagination, a wide range of feeders can be created, using readily available materials, such as the bottle-top feeder (**BOTTOM LEFT**) shown with a Great Spotted Woodpecker.

Fanciful and outrageous bird feeders

Over the years some very entertaining designs have been created, such as a bird bath that looks like a car wash. This section has a selection of designs that are both fun as well as practical.

*Feeder designs range from the basic and practical to those designed to blend in with a particular garden. In a small town garden, a pergola-style feeder (**LEFT**) might be appropriate: these are available from larger DIY stores, garden centres, and nature centres. As birds lack the aesthetic values of humans, they will not actually distinguish between the practical, whimsical or outrageous, so feeders are something you can let your imagination loose on. All the birds care about is that the food is kept fresh and dry. A modern town garden would be ideal for the hands (**BELOW**).*

ABOVE: *An example of an imaginative but essentially sensible design for a feeder. The cup-and-saucer has all the right elements. The cup contains the food supply, the saucer catches the waste – as many birds are messy feeders – and the cover prevents the cup becoming waterlogged, and spoiling the food. This idea can be modified – a chipped favourite mug or cup can be filled with suet and hung by the handle, for example.*

Section 3
Water

A constant supply of freshwater will also almost certainly guarantee a stream of visiting birds, particularly in suburban gardens, where water may not be readily available in summer.

ABOVE: *An ornamental bird bath is often a feature of a garden, but it does have a very useful function as well. If it is made from concrete, the inside should be sealed to slow down evaporation and, in winter, it will need to be kept free of ice.* **OPPOSITE:** *Wading birds, such as this Snipe, are rarities in most gardens, but if you have a large pond and ensure that the edges are shallow, migrating waders may stop for a few hours to feed.*

Why do birds need water?

Birds need water for two principal reasons, drinking and bathing. Like many other animals, some birds can metabolize a certain amount of water from their food. The digestion of carbohydrates such as seeds and grains releases water and many of the foods they eat, such as fruits and insects, contain a high proportion of water. But nearly all species need to take in additional fresh water. Perhaps the most remarkable of all the adaptations of birds to water, is that of the sandgrouse. They are a group of birds related to pigeons, mostly found in arid steppe and desert habitats, although in the past they have occasionally migrated as far as Britain and even bred. When rearing their young, the adult sandgrouse may need to travel several miles from the nest site to drink water. They also soak the plumage of their underside in water, before flying back to the nest, for the chicks to drink the water trapped in their feathers.

Birds also rely on having their feathers in tip-top condition. This both enables them to be able to fly, and also to keep warm in cold weather. One of the ways of keeping feathers in good condition is by bathing, and birds bathe in water (*see p69*) and also in dust, and with ants. Bathing not only helps keep the plumage clean, it may also help rid the bird of some parasites.

ABOVE: *A very wet House Sparrow.*
LEFT: *An adult and young Starlings bathing. Although birds may seem to simply enjoy a splash, bathing is important to keep their plumage in trim. Many birds also bathe in dust, and some bathe in ants' nests. Anting and dust bathing are believed to help control the parasites that often infest a bird's plumage.*

Supplying water

Water can be supplied in a wide range of ways – a pond is ideal. If you have a pond, make sure it is accessible to birds by having an extensive shallow edge, preferably with twigs and branches in the margin, rocks and bricks in the shallows. If you lack a pond – or as an addition to a pond – a bird bath on a pedestal is a good idea. This allows birds to keep an eye out for predators that might be attracted by their frenzied bathing activity. A bird bath should be wide and shallow, and will need cleaning regularly, both to ensure there are no build ups of algae (some of which can be poisonous), and also to stop it becoming a breeding ground for mosquitoes. If you can manage it, making the water move is always a good way of attracting birds, particularly during the migration period, when birds not familiar with the area might be passing through. Water can be made to move in a pond by the introduction of some sort of fountain, or waterfall. In a bird bath on a pedestal or on the ground, a simple drip can be all that is needed. A 25-litre tank with a clip controlling the flow from a short length of hose is all that is needed.

Finally, if you have only a very small garden, you can always provide water in a poultry drinker. These are available from any large pet shop or agricultural supplier, and contain anything from 1–25 litres. And most pet shops sell drinkers for canaries and other cagebirds. These are all better than a small bowl, because if you use a dish or bowl, there is a good chance it will dry up when birds most need it in

BELOW: *An idealized pond – large, with ample margins and shallows where birds can feed, and also areas of cover where other wildlife can hide. The centre of the pond should be fairly deep (more than 1m) to discourage marginals invading the whole pond. But do avoid planting with exotic water plants.*

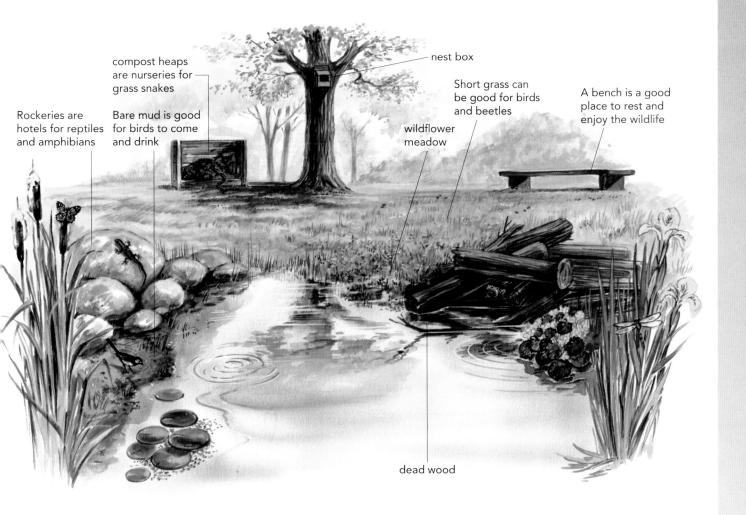

compost heaps are nurseries for grass snakes

nest box

Short grass can be good for birds and beetles

A bench is a good place to rest and enjoy the wildlife

Rockeries are hotels for reptiles and amphibians

Bare mud is good for birds to come and drink

wildflower meadow

dead wood

hot weather, or a single bathing session will more or less empty it.

In winter, it is often believed that birds need water, but most species can, and will, eat snow if it is available.

However, in dry conditions water is still needed, and a small aquarium heater, set at a very low temperature, is usually enough to keep a water supply from freezing. Anti-freeze, as used in cars, should NEVER be used.

LEFT: *a Bullfinch at a bird bath. This bird bath has been supplied with a small aquarium heater, which can be used to keep the water free of ice. The thermostat should be set as low as possible, and a timer can be used to switch it off during the hours of darkness, since nearly all the birds likely to visit are diurnal.*

BELOW: *A bird bath can also be suspended – a bird does not share a human's sense of aesthetics, so as long as you find something that serves the function well, use it. Even an up-turned dustbin lid will serve.*

Drinking water and bird baths

Water is an essential part of any plan to attract birds to a garden. Bird ringers have known this for years – when trapping birds to ring them for scientific studies, water is one of the best lures for a wide range of species – particularly if it is moving. A slight ripple will attract birds flying high overhead. This is a fact the bird trappers of old were familiar with, and used to lure birds flying past.

The classic bird bath is made from concrete or marble, and is a shallow dish on a column, or held up by a statue. But an upturned dustbin lid will do just as well, as will a large plant dish, and the birds won't notice the difference. If you have a pond, provide inviting and easy entrance and exit points for birds and other wildlife by ensuring that at least one side is shallow. Gently sloping twigs and planks with one end placed into the water will make convenient perches. Water features are popular in gardens, and will provide movement, but just as effective for attracting birds is a slow drip from a water tank. This can easily be constructed using a water butt, a length of plastic tubing, and a small clamp to restrict the flow to a regular drip.

Water can also be provided using commercial drinkers used by poultry farmers. Such drinkers are often available in larger pet stores; they have the advantage of slowing evaporation and keeping the water clean. But the birds cannot bathe in them.

ABOVE: *Custom-made bird baths are available from garden centres, nature centres and DIY suppliers, in a wide range of materials including cast aluminium (shown here), cast iron, concrete, marble and plastic. Cost will vary according to materials and design.*
LEFT: *One of the cheaper forms of bird bath on the market, but perfectly good for a range of birds. It is easy to keep clean, and shallow enough for bathing, as well as providing drinking water.*

Larger bodies of water

For the bird gardener, the most obvious use of a large pond is to attract waterbirds. Moorhens, ducks and other waterbirds will soon visit gardens with large ponds. Additionally, if the pond has a margin of rushes, and other aquatic and marginal plants, there is always the possibility of Sedge Warblers, Water Rails and other rarer species. Herons, even in towns, are often regular visitors, and may be unwelcome if they start feeding on pet goldfish or Koi Carp. But they are generally easy to keep at bay (see p39–40).

Most of us would love to encourage Kingfishers. While a large garden with a small river flowing through it is the most likely habitat to attract them, even a small garden with a large pond stocked with small fish as a food supply may entice the wanderers in autumn, or during severe cold weather. If you do have a large pond, or are on the side of a river, try to provide an embankment where Kingfishers (and Sand Martins) can tunnel, and with luck they may nest.

The shallow edge of a pond also provides a feeding ground for transient migrating waders, such as Snipes and Sandpipers, and perhaps more importantly it is a source of soft mud for House Martins and Swallows when they are nest building.

Because shallow water provides breeding grounds for mosquitoes and other small flies, ponds and waterways will attract Swallows and martins, which can often be seen swooping above the surface as they catch their airborne prey, and sometimes quickly dipping to sip a drop of water or to bathe.

ABOVE: *A Water Rail – not the most common of garden birds, but in rural areas, particularly in very cold weather, they may turn up.*

LEFT: *One of the most common waterbirds in larger gardens with ponds, the Moorhen is sometimes considered a pest as it will often attack other species of water fowl. They will feed mostly on small aquatic life, but will also take grain and poultry foods.*

RIGHT: *Kingfishers are probably top of every gardener's wish list for a bird to visit their garden. In autumn, particularly if you have a pond or a stream, they do turn up occasionally.*

BELOW: *House Martins do not visit feeders, but if they are nesting on your house or in the vicinity, you can help them by providing a shallow slope to a garden pond, or even shallow muddy puddles, so that they can gather the mud with which to build their nests.*

Section 4
Attracting More Unusual Birds

If the conditions and location are right, then unusual birds will come to feeders, often mixed with more common species. The best time is usually during the autumn migration season, or during periods of exceptionally cold, snowy weather.

ABOVE: *Waxwings periodically come to Britain, and feed on berries, particularly Pyracantha and others that are still available in late winter.* **OPPOSITE:** *The Little Owl is found only in larger gardens, often close to parkland; it is largely insectivorous.*

From time to time most land birds will visit garden feeders. Clearly waders, seabirds and many birds of prey are going to be much more difficult, if not impossible, to attract to a garden but there are still a huge variety that can be encouraged. The species you can attract depends very much on the size of the garden. But always remember that, as far as a bird is concerned, your garden is just a part of a wider countryside, a fragment of a much larger habitat, and birds rarely distinguish between the 'natural' and the man-made. This is why Kestrels and Peregrines will nest on ledges on buildings, and why many birds take to nestboxes and even more species will feed from the most unlikely containers. They are more interested in the food than its packaging. The most likely times of the year for seeing unusual birds are spring and autumn, the migration periods. In autumn, there is the additional factor that populations of bird species that do not migrate in the accepted sense often disperse. This is when juvenile Kingfishers turn up on garden ponds, for instance. Another time for seeing unusual birds is during prolonged hard weather, when the ground is frozen, or covered in snow. This is when many species of bird have great difficulty in finding enough food. At such times, bird feeders play an important role in the survival of many common garden birds and occasionally provide extra excitement for the birdwatcher when birds such as Waxwings turn up.

For the more unusual species, more unusual foods are sometimes needed. But water is also very important, and rare migrant birds passing through an area can often be lured with water, particularly if it is moving (*see page 70*). If you have the space, a pond with wide shallow margins can attract wading birds such as snipe or sandpipers. They may stay for only a few hours before moving on – but can be really exciting to see.

ABOVE: *A Kestrel in its nesting box. Although it is often difficult to attract birds such as Kestrels with food, providing a nesting place will often encourage them, even in cities and towns.*
LEFT: *Red-legged Partridges normally only feed on the ground, but they often become quite tame when they are not persecuted, and like Pheasants, will take food from a table despite their size.*

Birds of prey

The most common day-flying (diurnal) birds of prey in Britain, and most of northern Europe, are Kestrels and Sparrowhawks. Kestrels feed mostly on insects and voles, but also on birds and other small animals, preferring to hunt over open grassland, waste ground and similar habitats. For this reason they are only normally likely to hunt in larger gardens, and even then not on a daily basis. The best ways of attracting them would be to ensure that fairly large areas of grass are left unmown, and they can then hunt for voles, mice and insects – their preferred prey.

Sparrowhawks are another matter. Although during the 1960s their populations crashed (as a result of indiscriminate use of pesticides), they are once more widespread and often common. As their name implies,

ABOVE: *A Kestrel hovers in search of prey – and if you have a large garden and leave the grass long, this will provide habitat for grasshoppers and voles, both of which provide food for Kestrels.* **LEFT:** *Despite declining to the point of extinction in many parts of Britain in the 1960s, the Sparrowhawk is once again a common predator, and it is often necessary to take steps to ensure that birds feeding at a bird table do not become easy prey for them.*

their principal food is small birds, and many an enthusiastic garden bird watcher has been alarmed to see a Blue or Great Tit snatched from a feeder a few feet away. In fact the RSPB has received letters suggesting that protection should be removed from Sparrowhawks because of the damage they are doing to garden birds. They can indeed cause damage – I once had a dovecote with over a dozen white fantails, which over a period of a few weeks were reduced to a single pair by a large female Sparrowhawk. However, doves and tits are both part of the natural prey of Sparrowhawks and, despite appearances, they are unlikely to be having a lasting impact on the populations of wild birds. As mentioned elsewhere, the half a dozen Blue Tits you see at any one time at a garden feeder, may well be 100 or more different birds visiting in succession, as they constantly move round a neighbourhood. But if you think Sparrowhawks are taking too big a toll on the birds at your feeder, then you should reconsider the siting of the feeder.

A relative newcomer to bird tables is also spectacular – the Red Kite. A quarter of a century ago, it was teetering on the brink of extinction in

shops sell food suitable for birds of prey in the form of frozen day-old chicks, mice and rats, and there is no reason why owls should not be attracted to a night-time feeder stocked with prey. In parts of southern Europe and Africa feeding tables have been created for endangered vultures, with cattle carcasses, and large road kills gathered for them. In other parts of the world eagles, such as Sea Eagles or Bald Eagles, also come to feeding stations. In Asia the Parsee sky burials for humans are in reality large bird tables for vultures, and in the UK the Norfolk woodhenge, discovered in 1999, was possibly part of a site for feeding scavenging birds and other wild animals.

ABOVE: *The magnificent Red Kite, with its characteristic forked tail, can once again be seen soaring over many parts of Britain, and also comes to bird tables for household scraps such as the remains of chicken carcasses.*

BELOW: *A Tawny Owl requires a much larger territory than a normal garden in order to feed itself, but it can be encouraged by providing a suitable box for roosting and nesting in. Food, such as voles and mice, is encouraged by having large bramble patches and log piles.*

Britain, but a very successful reintroduction programme, followed up by the natural spread of wild-bred birds, means they are now becoming locally common again. Red Kites are primarily scavengers, and they can be attracted to gardens by putting out road kills, as well as household scraps such as chicken carcasses and other bones. At one feeding site in mid-Wales over 300 Red Kites gather in winter to be fed, and have now become a tourist attraction.

Owls are much more difficult to attract to a bird feeder, as they take almost exclusively live prey. However, I did once have a wild Tawny Owl fly into an aviary and take a dead chick that was intended for a tame owl. Many pet

Ducks and waterbirds

Clearly, in order to attract ducks and waterbirds, a pond is essential. Unless you have a large garden, with space for a large pond, then birds such as Moorhens and ducks will probably not be welcome – they can do a fair amount of damage to other wildlife, such as amphibians. However, if you are lucky enough to have a large pond, feeding ducks with grain and other commercial poultry feeds soon makes them tame –

then the problem often becomes how do you stop the numbers of Mallard and Moorhens getting out of control? Mallards can pollute the water of smaller ponds, and Moorhens predate other nesting birds and their eggs. However, a large pond, with a shallow edge on at least one side, with clumps of vegetation such as sedge, may also attract passing sandpipers, or in hard weather a snipe or two, to feed for a few hours.

RIGHT: *A Common Sandpiper – a bird not normally associated with gardens. However, particularly during the late summer and autumn migration period, they can be attracted to larger ponds, and may stop to feed if there is a wide muddy margin.* **BELOW:** *Mallards (the wild ancestor of the farmyard duck) are attractive birds, but should really only be encouraged in larger gardens. They will readily take grain or poultry food, but their droppings, and the 'puddling' from their feet makes a mess that is not usually acceptable in a small garden.*

Warblers

Warblers are a large and diverse group of small birds, some of which live in rather specialized habitats, such as reedbeds. They are mostly summer migrants, but some, notably the Chiffchaff and Blackcap winter in small numbers in southern England. All warblers are primarily insectivorous, gleaning insects and spiders in thick vegetation, often in the canopy. Particularly during the autumn migration period, almost any warbler species can turn up in a garden, and several, including Garden Warbler, Blackcap, Whitethroat and Lesser Whitethroat are particularly fond of blackberries and other soft fruits. Thick hedges, coppiced willow, and ivy are all good for attracting warblers. Closely related to the warblers are the crests – both Goldcrest and the rarer Firecrest feed among conifers and ivy. In fact, the infamous *leylandii* cypress provides good habitat for them. The Goldcrest often hovers like a hummingbird when gleaning insects and spiders from foliage.

LEFT: *The Firecrest is a very close relative of warblers such as the Chiffchaff, and like most warblers feeds mostly by gleaning insects. During the migration season, particularly in autumn, almost any species of warbler can turn up in a garden, where brambles, with their profuse, insect-attracting flowers, as well as their berries, are a good source of food.*

BELOW: *A fruit cage is often essential if you want to enjoy the results of your gardening. You can have the mesh large enough to keep out greedy larger birds such as Blackbirds and Starlings, but allow warblers and other small birds to gain access and feed.*

Flycatchers, wagtails and pipits

These are all almost entirely insectivorous birds, and for that reason relatively difficult to attract to gardens with food. Pipits and wagtails like open spaces – lawns and paved areas around ponds will often attract them, particularly the Pied Wagtail, which is often found in cities and towns; in fact the Pied Wagtail seems most abundant in man-made environments. Flycatchers will use nestboxes, the easiest way of attracting them to a garden. Flycatchers like to hunt from a prominent perch such as a tall post, or a telephone cable. If you are lucky enough to have a river or stream running through your garden, then a Grey Wagtail is a beautiful possible visitor, particularly in winter. But in all cases, an abundant supply of insects is essential – all too unusual in these days when many gardeners use proportionally more pesticides than an industrial farmer.

LEFT: *Pied Wagtails will feed on lawns and patios, particularly in an organic garden, where there are plenty of insects. Paths mown in long grass are ideal feeding areas for them.*

RIGHT: *Grey Wagtails are particularly attractive, with long bobbing tails. They are nearly always found close to running water, so most likely to be attracted to a garden with a stream or river close by, but they may also feed on a large pond, snatching insects from among the lily pads.*

Unusual finches and buntings

Crossbills are an eruptive species, with invasions from Scandinavia into the UK occurring sporadically. When these occur there is always the possibility of a small flock visiting a garden, particularly if conifers are present. In the recent past, the Reed Bunting was an occasional visitor to larger rural gardens, but it has become much more uncommon. Similarly, the Tree Sparrow has declined dramatically and is now a rare bird in gardens. The rarer buntings, finches and sparrows are all attracted by the same sorts of foods that are used for Goldfinches, Greenfinches and Yellowhammers (a bunting), and House Sparrows.

RIGHT: *Although a real rarity in most parts of Britain, where Crossbills do occur, pine trees are the only sure way of attracting them, as they are highly specialized feeders, extracting the kernels from pine cones.*

ABOVE AND RIGHT: *Yellowhammers, in common with a number of other birds, are polymorphic: that is to say they have several different plumages, with that of the breeding male being the most brightly coloured.*

Real rarities

Hoopoes were once common garden birds in southern Europe, and they occurred sporadically in southern England, occasionally breeding. Even now, they do turn up from time to time and are most likely to be seen feeding on a lawn or probing in a drystone wall, and a garden is as good a place as anywhere else. In autumn and winter, a more likely 'exotic' rarity, particularly in the east of Britain, is the Waxwing. Every few years their breeding population in Scandinavia and Siberia 'erupts' and they migrate to Britain, sometimes in flocks of a hundred or more. They feed on berries, in particular those species that are still available in late winter, such as firethorn. Supermarket car parks are often planted with shrubs and bushes, and indeed Waxwings have been seen in such unlikely places. Fifty years ago Wrynecks still occurred in gardens in southern England, where they are now extinct.

ABOVE: *In Britain the Hoopoe is a real rarity, although it has bred from time to time. In southern Europe Hoopoes are not uncommon in gardens where they feed on lizards as well as large insects.* **LEFT:** *The Bluethroat is widespread in Scandinavia with scattered populations in other parts of Europe. It turns up occasionally in Britain on migration, when it is often rather drab, as only the breeding male has the spectacular plumage from which it gets its name.* **BELOW:** *When Waxwings 'invade' Britain from Scandinavia, they are as likely to turn up in gardens as anywhere else, and if they find a good supply of food will often continue to return for several days.*

Escapees

Escaped cagebirds are often attracted to bird tables – they are accustomed to being fed, and often find it difficult to fend for themselves in the wild. Canaries and Budgerigars are the commonest – largely because they are common in captivity. The Rose-ringed Parakeet has successfully colonized a large area of London and its suburbs and a few other parts of England, and continues to spread. In winter it is largely dependent on feeders, particularly peanuts, for its survival. Other common escapes include Cockatiels, Zebra Finches and lovebirds – but almost anything that is kept in captivity can, and often does escape, from eagles to flamingos.

LEFT: *A Rose-ringed or Ring-necked Parakeet at a feeder. Although they are largely dependent on being fed during the winter months, they are very hardy birds, and their native range extends way into the Himalayas.*

Gardens as nature reserves

Some of the birds coming to feeders, or using gardens to breed in are now considered to be of conservation concern. Endangered birds, or birds that have shown sudden declines in recent years, are 'Red Listed'; those not so threatened or with stable populations are 'Amber Listed'; those not considered in any immediate danger are 'Green Listed'.

Surprisingly, a few Red Listed species are garden birds, showing how important bird feeding can be. These include

Song Thrush
Marsh Tit
Willow Tit
Starling
House Sparrow
Tree Sparrow
Bullfinch
Yellowhammer

Among the 'Amber Listed' species that are found in gardens and may benefit from feeding are the following species:

Green Woodpecker
Dunnock
Blackbird
Fieldfare
Redwing
Mistle Thrush
Goldcrest

Other Red or Amber List species which are often seen in gardens but rarely visit feeders (though they will certainly benefit from increased insect populations) include:

Swallow
Spotted Flycatcher

TOP RIGHT: *Song Thrush*
CENTRE RIGHT: *Bullfinch*
BOTTOM RIGHT: *Goldfinches.*
Monitoring the numbers of common birds is one way that non-expert enthusiasts can help scientific research into long-term trends, and both the Bullfinch and the Song Thrush have been 'red listed' because of the dramatic decline in their numbers.

Section 5

Directory of Bird Foods

There are literally thousands of plant species in cultivation, and there are hundreds of foods sold in supermarkets that are suitable as bird foods, and there are hundreds more that are available as seeds from specialist suppliers that can be grown in a garden. This directory contains a selection.

ABOVE: *Goldfinches feed on a wide range of small seeds particularly thistles and similar species, such as teasels. These are attractive architectural plants that will enhance a garden, as well as providing food for Goldfinches.* **OPPOSITE:** *A female blackbird – a common sight feeding on berries in the winter.*

Grains and seeds

The simplest of all foods for feeding birds are dried grains and seeds, as they are easy to obtain and easy to store. And because they are the food store from which a plant develops, they have a high nutritive value for birds. Most of the major suppliers of wild bird foods now produce catalogues, with a considerable amount of information on the foods, and they also have a wide range of mixtures for use in appropriate feeders.

It is important to store dried bird food well. The food must be kept dry and in containers that cannot be attacked by pests – though in the case of bird foods this is not so important, as they may well find mealworms and weevil larvae more attractive than the seed they are attacking.

Oats

Avena sativa

Description: Like many other grains this is an annual grass, but unlike barley, wheat and rye the grains grow on open, spreading panicles. The wild oat *A. fatua* is a

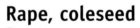

weed of cereal crops, and can often be seen in wheat and barley crops, as it is usually taller.

Food value: Sold in a wide range of forms – rolled (porridge), clipped, groats (hulled grain) and oatmeal as well as whole grains. All have a relatively high energy value – it has a fat content higher than most other grains.

Birds: Coarse oatmeal (pinhead oat) is a useful grain for bird feeding as even small birds will take it. Pigeons and doves eat the whole grain, as do ducks and Moorhens.

Rape, coleseed

Brassica napus

Description: Derived from the wild turnip, rape is a member of the

cabbage family, Cruciferae. The vegetable swede and rape are considered to be varieties of the same species with differing root formations. It is widely cultivated in vast monocultures and the oil has a wide range of culinary and industrial uses. It is an annual, growing to about 1 m high, with a spike of bright yellow flowers, like small wallflowers, followed by pods of tiny black or brown seeds. It is easily grown in the garden, where it will also attract some of the cabbage-loving butterflies. Mustards are very closely related and very similar in appearance. In fact, the classification of mustards, turnips, swedes, rapes and kales is a taxonomic nightmare.

Food value: The seeds have a high oil and protein content. The 'mustard' in mustard and cress salads is often replaced with rape seed which is less pungent, but easier to harvest. Rape seed contains about 40% fatty oils, and 22% protein.

Birds: Used in mixtures for a wide variety of seed-eating birds.

ABOVE: *Linnets are birds that love hedges and rough grasses, and will also be attracted to a range of seeds and grains put out for them.*

Hemp, marijuana
Cannabis sativa

Description: A tall tender, annual herb, growing up to 4 m tall. The stems and leaves are covered in fine hair; the leaves are spear-shaped with toothed edges, and the male flowers are in small yellowish green panicles, less than 5 mm in diameter; the female flowers are darker green and more tightly bunched. The seeds are small, shiny and dark brown. There are two forms of this species: one is grown for its fibres, and produces relatively little of the narcotic drug for which it is famous; the other, which produces a narcotic resin, is grown in warmer climates, often illegally.

Food value: High in oils and proteins, it has long been a constituent of cagebird seed mixes, and is still widely available for this purpose.

Birds: Finches, especially Linnets and Chaffinches.

Safflower
Carthamus tinctorius

Description: A thistle-like annual, growing up to nearly 1.5 m. The stem is branched and spiny, and the leaves dark green with spiny edges. The flowers are thistle-like, with a tuft of bright orange-yellow petals. The seeds are used to extract oil, which has a wide range of culinary and industrial uses. It is not known in the wild but it is assumed to have originated in the Middle East and is now widely cultivated in Egypt, Ethiopia and Afghanistan, as well as other countries. It grows wild in areas around the Mediterranean as a result of obsolete cultivation.

Food value: The seeds have a high energy value – over 500 calories/100 g, and are an excellent source of vitamins A and B6, thiamin and several minerals.

Birds: Goldfinches, Chaffinches, Siskins and other small seed-eaters are all attracted to the seeds. *The seeds are bitter and squirrels and other mammals do not normally eat them.*

Quinoa
Chenopodium quinoa

Description: It is one of relatively few grains that is not a grass.

Food value: Highly nutritious, and a good source of energy; it also has a very high iron content, as well as other minerals.

Birds: Almost all small seed-eating birds will eat it, as well as tits. Other species such as Robins will eat it cooked.

Pumpkin
Cucurbita pepo

Description: The vegetable marrow family is a large one, probably originating in the Americas, and the pumpkin is one of the largest, growing to weights of over 100 kg (the British record is 352 kg (774 lb), the world record nearly 682 kg (1500 lb)). It is closely related to cucumbers, courgettes (zucchini), marrows and squashes. Most are climbing or trailing annuals, that produce large yellow flowers, and large fleshy fruits.

Food value: The flesh is rather watery, and it is the seeds which are normally used as bird food. They are similar in size to sunflowers seeds, thinner and creamy-white, and are rich in protein and fats. The watery flesh has 26 calories/100 g, but the seeds have over 540 calories/100 g, which includes 46% fat, 18% carbohydrate and 25% protein, and is very high in iron, some other minerals and vitamin A.

Birds: Tits, Greenfinches and almost any species that will take sunflower seeds will take the unshelled seeds. A much wider range will take the kernel if they are shelled.

Buckwheat
Fagopyrum esculentum

Description: An annual growing to about 65 cm, with hollow stems, tinged pink or purple, and arrow-shaped leaves. The flowers grow in panicles, and these are followed by small brown grains. Although it grows wild in some parts of Britain, it is mostly known as a cultivated plant. Its grain has become popular for culinary purposes, and is widely available in supermarkets.

Food value: A good source of energy, containing about 24% carbohydrate, and 13% protein, as well as some calcium and iron.

Birds: Popular with most small finches, sparrows and buntings.

BELOW: *Snow Buntings are unlikely to be found in any but the largest gardens. However, on a nature reserve, they can be fed with a wide range of seeds, and like most finches and buntings prefer the smaller seeds.*

LEFT: *Pheasants are attractive, but large and greedy, taking a wide range of seeds and grains. They are also predators of small amphibians, lizards and other wildlife.*

Lawn grasses
Gramineae spp.

Description: Several species of grass are used for lawns and pastures. These include rye grasses (*Lolium* spp), meadow grasses (*Poa* spp) and fescues (*Festuca* spp.) Left uncut the stems grow to between 30–100 cm, and they produce abundant seeds which are eaten by many birds. Seed can be purchased from agricultural suppliers as, probably because of its relatively high price, bird food suppliers do not often offer it. Long grass is also an important source of insects, such as grasshoppers. When mowing it is always best to set the blades as high as possible, and minimize the damage to insects, amphibians and reptiles.
Food value: Like all seeds they have high energy value.
Birds: Many small birds, but particularly finches and sparrows.

Nyjer, niger, noug
Guizotia abyssinica

Description: An annual, growing to about 1.5 m tall, in the aster family. Related to the Michaelmas daisies, it is indigenous to Ethiopia where it is grown in rotation with cereals and pulses. It is an important oilseed crop, supplying 50% of Ethiopian oilseed production.
Food value: Cagebird keepers have long used it as a substitute for thistle seeds, and it is now used extensively for feeding wild birds. It can be scattered on the ground or on a table, or supplied in a Perspex™ tube feeder.
Birds: The preferred food of Goldfinches, it also attracts a wide range of other small birds. It can be used in tube feeders, on bird tables or on the ground.

Sunflower

Helianthus annuus

Description: Sunflowers are fast-growing annuals, often reaching 2–3 m or more in height, with large yellow flowers up to 30 cm in diameter. A familiar garden flower, it probably originated in Mexico. It is cultivated for its seeds, which produce a high-quality oil used in cooking, margarine, soaps and other processes. When grown as a garden plant the flowers should be left, to allow the seeds to develop, and birds will feed on them in situ.

Food value: The flowers are very attractive to insects. The seeds are available in two principal forms: striped and black. Black sunflower is the best, with a very high energy content (it is the type grown for oil production) and it is this type that is recommended for feeding birds. The striped varieties may be cheaper, but they are less nutritious. Sunflower seed can be supplied in tube feeders, on tables or on the ground. Not all birds can open the seeds, and sunflower hearts (kernels) can also be purchased. The kernels are 50% fat, 19% carbohydrate, 23% protein and 3% sugars.

Birds: Sunflower seeds are eaten by a very wide range of birds; tits, woodpeckers, Nuthatches and finches can all deal with them in their shells, and other birds such as Robins and Dunnocks can eat the hearts.

Barley

Hordeum distichon

Description: An annual grass similar to wheat, but always awned (see Rye, p94). When it is ripe the ears droop. Like wheat it probably originated in western Asia, and is now grown throughout the temperate regions of the world.

Food value: Most barley is grown for making malt for use in the beer brewing industry. Barley that is not up to standard is sold as animal feed. It is one of the most nutritious grains, being high in carbohydrates and protein, and a good source of thiamin (vitamin B1).

Birds: As for wheat and other grains, most useful for larger birds, if bought in bulk.

RIGHT: *A Coal Tit takes whole black sunflower seeds from the feeder and opens them on a nearby perch.*

Linseed, common flax

Linum usitatissimum

Description: It is an annual herb, growing to about 1.5 m tall, with clusters of bright blue five-petalled flowers. The seeds are roughly spherical, slightly flattened, less than 6 mm long, brownish and very shiny. The stems of the varieties grown in the north are used for making linen, the seeds grown in warmer areas are pressed for linseed oil, and the residue is used in linseed cake as animal feed.

Food value: It is a high-energy food (nearly 500 calories/100 g), rich in fatty oils (34%) and protein (20%), and with several minerals present.

Birds: The seed is commonly used in the mixtures used for cagebirds, and also wild bird mixes. Most small seed-eating birds including sparrows and finches thrive on it, and even some small insectivorous birds such as Robins will take it.

PUTTY EATING

Members of the tit family frequently eat the putty around windows, particularly in late autumn and winter. They may have a taste for the linseed in the putty, or it is possible that putty contains minerals, such as calcium, or other substances lacking in the bird's diet and it is trying to make up for the deficiency. Other acts of vandalism such as paper-tearing may be connected. It can be difficult to deter this behaviour, but you can try brushing the putty with aluminium ammonium sulphate, which is distasteful and may help stop the habit.

Great Tits and Blue Tits are often credited with considerable intelligence, as they learned to peck through the foil tops of bottled milk in order to feed on the cream that gathered at the top.

RIGHT: *Whinchats are Robin-sized members of the thrush family. They are insectivores, and mostly found in open heathland habitats, but may occasionally be encountered in larger gardens while on migration.*

Rice

Oryza sativa

Description: Originating in Asia, like most other cultivated cereals, rice is grown in a very wide range of widely differing varieties. There are two principal groups, one grown in water, the other grown in drier conditions. It is mostly grown in tropical Asian countries, but some is grown in the Mediterranean climates, and the world's largest exporter is the USA. The types sold vary in the degree of 'cleaning'. After the husk is removed there is a brown or reddish bran layer, which is usually removed to produced a pearled or polished rice. 'Wild rice' *Zizania aquatica* is a grass native to North America.

Food value: High in starch, but fewer proteins than other cereals, particularly if polished and cooked.

Birds: Can be used either as a grain, which finches, doves and other birds will eat (and can be purchased very cheaply in supermarkets), or cooked, when

other species, including Robins and Dunnocks will take it, but with its nutrition value dramatically reduced.

Millet

Panicum miliaceum

Description: A tall grass, grown as a staple crop in many parts of Africa and Asia, eaten as a cereal or used in brewing beer. It is grown as birdseed in other parts of the world. The seeds grow as dense clusters, and it was one of the earliest cereals to be cultivated, probably originating in the Middle East. There are several other species of millet, mostly grown as birdseed. They are not frost tolerant, although some species can be grown in warmer parts of Britain and Europe. The seeds are small and there are over 175,000 seeds in one kilogram. The foxtail millet *Setaria italica* is the type most commonly sold as whole sprays for cage and aviary birds.

LEFT: *Starlings are attractively plumaged birds, which, were they not so common, would be highly regarded. They are one of the most abundant birds at feeders, particularly in winter, when migrants from Scandinavia and Russia augment their numbers.*

Food value: A wide range of millets is available, as might be expected of a plant that is extensively cultivated; it is rich in protein and iron. Its energy value (378 calories/100 g) is comparable to most grains.

Birds: Although long popular with cagebird keepers, the seed heads of fox-tail millet do not seem particularly popular with wild birds. However, certain ground-feeding birds such as Chaffinches eat the loose millet seed.

THE RELATIVE NUTRITIONAL VALUE OF DIFFERENT TYPES OF RICE

Rice type	Calories/100 g	Total Carbohydrate %	Protein %
Long Grain Brown, raw	370	77	8
Long Grain Brown, cooked	111	23	3
White, polished, cooked	97	21	2

SOURCE: www.nutritiondata.com

Maw seed, blue maw, poppyseed

Papaver somniferum

Description: An annual originating in Asia that grows to about 75 cm, with greyish green leaves, and usually pink flowers, although there are many cultivated varieties. Its cultivation is illegal in many countries. After flowering large seed capsules are produced that contain huge numbers of very small seeds.

Food value: An important seed for feeding many small birds, it has been used by aviculturists, mixed with other foods, for insect-eating species.

Birds: Could be used mixed with other foods for Wrens, Dunnocks warblers as well as finches, such as Goldfinches and Linnets.

Canary seed, canarygrass

Phalaris canariensis

Description: This is an annual herb, growing to 100 cm. Over 200,000 tonnes are grown world-wide, almost exclusively for feeding birds – originally cagebirds, but now, increasingly, wild birds. Although the plant is originally from the Mediterranean, Canada is now a major producer.

Food value: Because of its high protein, high oil, and high unsaturated fat content (84%), it is an excellent food for small birds.

Birds: A very wide range of small birds will eat this seed, including insectivorous species.

Sorghum, kaffir corn, milo

Sorghum vulgare

Description: A tropical grain related to millet, which looks rather like maize while it is growing, but has a single large flowering spike. The seeds are round and either white or red. If it could be grown (by germinating in a greenhouse, and planting out when all chance of frost has passed) it would be very attractive in a garden.

Food value: Although widely eaten by humans, and used for brewing, in Africa and Asia, elsewhere it is mostly used in animal feeds. Its energy values are similar to other grains.

Birds: Suitable for most seed-eating birds.

LEFT: *Dunnocks are shy, skulking birds, and although they will visit bird tables, prefer feeding on the ground, usually close to shrubs and bushes.*

Rye

Secale cereale

Description: Similar to wheat, but capable of being grown in much colder climates. The ears are like wheat, but are 'awned' (awns are long hair-like bristles on each grain). Although some wheats are awned, most are not. Like wheat, the grain is threshed and winnowed to remove the husks and awns before being marketed.

Food value: Less carbohydrate (less than 25%) than most other grains, but high in protein and minerals.

Birds: As for wheat, mostly larger birds, including waterfowl, but finches and other small birds will feed on it.

BELOW: *If there is space in the garden, plant some cereals, such as wheat and maize – or almost any of the temperate species of grain described here. They make an unusual border to a vegetable patch, and provide food for birds.*

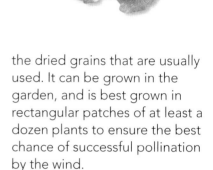

Wheat

Triticum aestivum

Description: One of the most widespread and common cereal crops, originating in western Asia, it is an annual grass growing in a wide range of varieties. The seed head grows on a hollow stem (straw). In modern times short straw varieties have become dominant, but in earlier times it grew to a metre or more. It can easily be grown in a garden, where mixed with other cereals and arable weeds such as cornflowers, corncockles and cornmarigolds, it makes an attractive summer display.

Food value: Highly nutritious, with around 75% carbohydrate content, as would be expected since it is a staple of the human diet in many parts of the world. It is usually sold as flour for human consumption.

Birds: A wide range of birds will take grain, but it is most popular with larger birds such as doves. Also fed to birds in the form of bread.

Maize, Indian corn, sweetcorn, corn-on-the-cob, mealies

Zea mays

Description: A large grass, originally from South America, maize grows in an astonishing range of colours, shapes and sizes, but the majority sold is in the familiar form of sweetcorn. Maize grows to about 2 m and each plant bears one or two cobs. Although for human consumption it is usually eaten fresh, for birds it is the dried grains that are usually used. It can be grown in the garden, and is best grown in rectangular patches of at least a dozen plants to ensure the best chance of successful pollination by the wind.

Food value: Maize is used for birds in three principal forms: whole grains, flaked and kibbled. It is an excellent source of carbohydrate (74%), but lower in proteins than other grains.

Birds: Mostly eaten by larger birds such as pigeons and doves, and Pheasants, but other species will take the smaller kibbled grains.

THE RELATIVE ENERGY (CALORIES) AND OTHER VALUES OF SOME OF THE COMMONEST GRAINS THAT CAN BE USED TO FEED BIRDS

Grain	Calories/100 g	Total Carbohydrate %	Protein %	Extras
Maize	365	74	9	Vitamin A
Buckwheat	343	71	13	Minerals
Wheat	342	76	11	—
Rye	335	23	15	Minerals
Barley	354	73	12	Thiamin
Oats	389	66	17	Thiamin; Iron
Quinoa	374	69	13	Minerals
Brown rice	370	77	8	

SOURCE: www.nutritiondata.com

Mixed seeds

Nearly all bird food suppliers sell mixed seeds for feeding birds. However, according to research in the USA, which is almost certainly applicable elsewhere, this is a very inefficient way of feeding birds; my own observations certainly bear out these findings. This is simply because birds nearly always take only their preferred food from a feeder, and discard other seeds and grains. When food is provided on the ground, this is not too wasteful, but when mixed seeds are used in elevated feeders, such as a tube feeder, a great deal of food is wasted. This is particularly true when grain is mixed with preferred seeds such as black sunflower. It is always better to have a single type of food in a feeder, and also to have that food in more than one type of feeder, preferably varying in height as well. Birds that feed on certain seeds and grains do not like feeding on the ground, while others do not like tube feeders. To quote Dr Aelred Geis: 'The failure of wild bird mixes to accommodate the ecological differences among bird species very seriously detracts from their effectiveness.' But as so often in the modern world, wastefulness is profitable, and seed merchants will continue to market mixtures to the uninformed public. Dr Geis continues: 'The striking differences in seed preferences among bird species clearly suggests why wild bird food mixes are inefficient and illogical. Unfortunately, the bird feeding public has been trained by the bird food industry to use wild bird food mixes. This usually results in lower feeding efficiency, wasted money, and other problems. *At every opportunity, an effort should be made to educate the public about the inappropriateness of mixes.*' [my italics].

BIRD CAKE RECIPE

Take a suitable container, such as a clean flowerpot. Put a wire through the middle, to hang it by. Fill it with a mixture of melted suet* (not lard), mixed seeds, and finely chopped, dried fruit, in the ratio of 1:3. Make sure it does not contain anything that might go rotten, but dried bread or cake crumbs could be included. When the suet is hardened, hang it like a bell – the flower pot will keep the contents dry, and ensure that only the more agile birds can feed at it.

*Suet is a relatively hard fat that grows around the kidney of cattle. It is not to be confused with lard or dripping, which melt easily and are very greasy

ABOVE: *Great Spotted Woodpeckers are attracted to fats, such as this bird cake, and over the past half century have increased greatly in suburban gardens, becoming regular visitors to most bird tables.*

Bird feeding do's and don'ts

- Do not put out mouldy food, as moulds can produce toxins that are fatal to birds.

- Do not put out food that could choke a bird – remember they cannot bite or chew – so long bacon rinds are potential hazards.

- Always put peanuts in a holder, particularly in the breeding season, when whole nuts could be fed to a nestling and choke it.

- Make sure drinking water is clean and changed regularly.

- Avoid salty items, such as salted peanuts, salted crisps and other snacks.

- Don't place feeders where the birds will be vulnerable to cats, collision with windows or with cars.

- Don't use mixed seeds in elevated feeders. Only use them on the ground but, better still, avoid mixes, and feed individual seed types.

RIGHT: *Placing a bird table in the open ensures that birds can see potential predators easily.*

Nuts and pulses

Like other seeds and grains, nuts and pulses are the food store for the developing seedling, and are therefore highly nutritious. Many types of this foodstuff can be purchased more cheaply in bulk from pet food suppliers, and a considerable range is available in health food shops. Pulses, such as dried peas and lentils, not only form part of our staple diet, but are also useful as bird foods. Nuts can be fed whole, in their shells, shelled or pulverised. They should all be stored in dry, reasonably airtight containers – bulk, plastic garbage bins are ideal.

Peanut
Arachis hypogaea

Description: A legume (member of the pea family) that grows its peas underground. The plant grows only in frost-free areas, and the 'nuts' are imported into the UK. They can be purchased in their shells, but more normally are bought shelled. It is important to buy good quality peanuts, as moulds such as aflatoxin can be fatal to birds.

Food value: Highly nutritious, rich in fat, carbohydrate (mostly sucrose) and protein. Also contains vitamin C, iron and calcium. With over 560 calories/100g, and nearly 50% fat content, peanuts are an excellent source of energy.

Birds: A very wide range indeed will be attracted by peanuts. Siskins, Greenfinches, Nuthatches, woodpeckers, and most tits. Peanut granules, which are claimed to be even higher in energy content, are taken by Robins, Dunnocks and other small birds. It is usually supplied to birds in mesh feeders, as otherwise most species will carry them off to eat, and there is also a risk of smaller birds attempting to feed a whole peanut to a nestling and choking it.

LEFT: *Peanuts are by far the most popular food for wild birds. They are best fed from a mesh container, so that the birds can be seen while feeding. But always ensure good quality nuts are used, to avoid aflatoxins, which are colourless and odourless.*

Chick pea

Cicer arietinum

Description: A spreading annual up to 50 cm high, covered with tiny club-shaped hairs, and with greenish white flowers. The pods contain a single pea, up to 10 mm in diameter, with the skin removed.

Food value: Although they have a similar energy content to lentils (around 350 calories/100 g), they have a lower protein content and are less nutritious.

Birds: Pigeons, Pheasants. Like other grains and pulses, it can be ground coarsely to make flour, which many other birds will eat.

Coconut

Cocos nucifera

Description: The large, meaty nut of the tropical coconut palm, growing to 24 m. It is normally sold without its even larger fibrous husk (coir). Coconuts are normally sawn in half and hung out on bird feeders. Other products derived from coconuts are also popular with birds, including shredded desiccated coconut.

Food value: A good all round food. The meat contains over 350 calories/100 g, and comprises 33% fat, 15% carbohydrate and 6% sugars. It is also a source of vitamin C, calcium and iron.

Birds: Coconut halves are very popular with most species of tit, and also Great Spotted Woodpecker. A wider range of birds, including Robins, takes desiccated coconut. However, desiccated coconut should always

be soaked thoroughly before being offered, as otherwise it may swell inside a bird's gut and kill it. Nesting birds can use the fibres from the raw coconut husk for nest building.

RIGHT: *A coconut sawn in half will provide not only a highly nutritious food for birds such as these Great and Blue Tits, but also entertainment for the watcher, as the birds hack away at the meat inside the hanging shell.*

Lentil
Lens culinaris

Description: One of the earliest of the pea family to be cultivated, it was first grown in the eastern Mediterranean or Near East, but can be grown as far north as England. Lentils are low-growing annuals, similar to peas, generally with greenish white flowers. Each seedpod contains only one or two seeds. They are normally sold dried, and there is a great range of size, as well as colour — greens, reds, browns, and yellows.

Food value: As most vegetarians know, lentils are a good source of protein. They are very nutritious, with a protein content of about 25% by weight.

Birds: Most seed-eating species will eat lentils, and if they are cooked, other species such as Robins will eat them.

RIGHT: *Collared Doves are relatively recent colonists of the British Isles, having arrived in the 1960s, but they are now a familiar sight in many rural and suburban areas, where they feed on grain and seeds.*

Mung bean, green gram
Phaseolus aureus

Description: Originally from India, it is a climbing annual, with purple flowers and rather hairy stems and leaves. The small beans are green, brown or mottled, and are most often used for growing bean sprouts. Each pod may contain up to 15 beans.

Food value: In addition to a high-energy content, they are a very good source of protein, and contain vitamins A and C.

Birds: Popular with doves and pigeons, and larger seed-eating species.

Pea
Poa sativum

Description: A climbing annual, it probably originated from the Near East, and was brought to Britain by the Romans; it now occurs in many varieties. The most usual is the garden pea (*var. arvense*), which has white flowers, and bright green seeds, and usually 4–10 peas in a pod. It is easily grown from seed, but mice often eat the seeds before they have an opportunity to germinate.

Food value: High-energy food, but not as nutritious as lentils.

Birds: Dried peas are a particular favourite of pigeons and doves.

Mixed nuts

Description: Mixed nuts are sold in supermarkets and these, as all nuts, make excellent bird food. They include: Brazil nuts *Bertholletia excelsa*, which are large, very fatty nuts with a high energy content, from South American rainforests; the walnut *Juglans regia*, which is native in south-east Europe across to Asia and is widely cultivated – the hard shells of the nuts contain the characteristic brain-shaped kernels; cashew nuts *Anacardium occidentale*, which come from a tropical tree, and grow in a strange way – the nuts hanging beneath a large, fleshy 'apple'; almonds *Prunus dulcis*, which can be grown in Britain, but are mostly grown around the Mediterranean and Middle East – the nut is very hard shelled. Other nuts (perhaps Christmas leftovers), which can be used as bird food include pecans *Carya illinoensis*, pistachios *Pistacia vera* and

chestnuts *Castanea sativa*. For the reasons given for seed above, mixed nuts are best fed on a platform or from a dish with drainage holes, so that birds can choose the nuts they like without discarding or wasting less preferred varieties.

THE RELATIVE ENERGY (CALORIES) AND OTHER VALUES OF SOME OF THE COMMONEST NUTS THAT CAN BE USED TO FEED BIRDS

Type of nut	Calories / 100 g	Fat %	Total Carbohydrate %	Protein %	Extras
Brazil nut	656	66	12	14	Rich in minerals
Walnut	654	65	14	15	Iron and calcium, and trace of vitamin C
Almond	578	51	20	21	Minerals and vitamin E
Cashew	566	47	27	25	Minerals
Chestnut	369	4	78	5	Vitamin C
Pecan	691	72	14	9	
Hazel	628	61	17	15	Minerals and vitamin C
Pine kernels	629	6	61	12	Vitamin C, thiamin

Plants, shrubs and trees

One of the best ways of feeding wild birds is to create 'natural' fruits and seeds that are attractive to them. A large number of plant species disperse their seed by being attractive to birds, so that birds eat them, and the seeds pass through the digestive system and are dropped away from the parent plant. All the plants described in this section are readily available in garden centres and nurseries. In addition, a small area in the garden can be sown with some of the seed from a wild bird mix. Sunflowers, flax and other annual species will grow, and make an attractive splash of colour in late summer.

Fat hen

Chenopodium album
There are many closely related species in the family Chenopodiacae which provide useful seeds for small birds, some of which are occasionally grown as crops, these include Good King Henry and orache.
Description: A medium-sized herbaceous plant, with dark green leaves, covered with greyish powdery meal, and a reddish stem. It is usually most common on recently disturbed ground. It is closely related to Good King Henry and the Goose-foots, as well as the beets and oraches – all of which are also food sources for birds.
Food value: The seeds are a valuable food source for many small birds.
Birds: Chaffinches, Greenfinches, sparrows and other small birds.

FLOWERING PLANTS THAT PRODUCE SEEDS AND BERRIES, WHICH PROVIDE FOOD FOR BIRDS

Teasel	*Dipsacus fullonum*	Large seed head, which is attractive to Goldfinches.
Honesty	*Lunaria biennis*	Self-seeds easily; attracts finches, including Bullfinches.
Dandelion	*Taraxacum officinale*	Can be grown as a cultivated salad; the seeds are eaten by Goldfinches.
Lavenders	*Lavendula* spp.	Fragrant flowers attract insects, and finches eat seeds.
Honeysuckles	*Lonicera* spp.	Fragrant flowers attract insects, and birds eat sweet, red berries.
Evening primrose	*Oenothera biennis*	Pretty flowers, followed by seeds for finches.
Greater plantain	*Plantago major*	A weed that looks good when cultivated, with seeds for doves and finches.
Fat hen	*Chenopodium album*	A common weed, seeds suitable for sparrows, finches and doves.
Golden rod	*Solidago virgaurea*	Thrives on poor soils, and produces seeds for Goldfinches, Siskins and Linnets.
Groundsel	*Senecio vulgaris*	A very common weed, with seeds eaten by finches and sparrows.
Cornflower	*Centaurea cyanus*	A brilliant splash of colour followed by seeds for finches and sparrows.
Thistles	*Cardus* spp.	Seeds eaten by Goldfinches and other finches.

BELOW: *Even the smallest garden can provide food for wild birds, with decorative plants such as Teasel being a favourite of Goldfinches.*

Alder
Alnus glutinosa

Description: One of several rather similar species of deciduous trees, growing to a height of 20 m, although it is often pollarded. Alders normally grow close to water and are often planted on river banks or by ponds and streams. The leaves are rounded, with a toothed edge. In late winter, before the leaves have appeared, the male's catkins are produced on the same stalk as the small brown flowers of the female. The blackish fruits are like small pine cones.

Food value: No data has been located for this species, but it is popular with several bird species.

Birds: Redpolls are specialist feeders on alder cones, though other birds also feed on them, including Siskins.

Hornbeam
Carpinus betulus

Description: Although superficially similar to the beech, it is actually more closely related to the hazel. It is often a very large tree (up to 25 m), and has a smooth greyish trunk. It produces greenish catkins in spring, and the small nuts grow within leafy bracts. Hornbeams are often grown as hedging plants, and provide good shelter as they will retain their leaves when clipped. They are also pollarded and coppiced, providing important habitats for birds and other wildlife.

Food value: No data has been located for this species, but it is popular with several species of bird.

Birds: The nuts or 'mast' produced by Hornbeam is an important food for several finches, including Brambling, Chaffinch, and the now extremely rare Hawfinch.

Hazelnut, cobnut, filbert

Corylus avellana

Description: A small, deciduous tree usually grown as a hedge or regularly coppiced. It has almost circular leaves, and in spring long dangling male catkins, up to 6.5 cm; the female flowers are separate, short erect spikes with red styles. The rounded nuts mature in late summer, but when squirrels and mice are present a bush can often be stripped before they can be harvested.

Food value: The nuts are rich in fatty oils and vitamins. They are popular as human food and tend to be relatively expensive.

Birds: The whole nuts, complete with shell, can be dealt with by only a few species, notably Nuthatches, which wedge the nut in a crevice and attack it with their bill. Shelled hazel nuts are popular with a wide range of birds, particularly tits.

Beech

Fagus sylvatica

Description: A large deciduous tree (up to 30 m) with a smooth, grey bark. The buds are brown, with bright green leaves, gradually darkening. The flowers are rather insignificant, and often escape notice – the male catkin is a short tassel below the female flower. The nuts (mast) have a husk with bristles, which opens to shed the mast. The mast is erratic, appearing in great quantities but only irregularly. A copper coloured beech is often grown in gardens.

Food value: A very high-energy food with over 500 calories/100 g, it also contains vitamin C. The canopy of the tree in leaf also provides a good hunting ground for warblers and flycatchers.

Birds: Beech mast is associated with Bramblings and Chaffinches, which sometimes congregate in large numbers in beech woods, often with tits and other species.

TREES

Strangely, none of the enterprising bird food suppliers seem to be offering wild trees, nuts and seeds. This is surprising since some of them produce seed in abundance. However, the keen bird feeder can always collect their own – ensuring it is legal to gather. Acorns, beech mast, alder cones, and even pine nuts would all make a welcome addition to the bird table that could easily be gathered from the wild and stored for later use.

Pedunculate oak

Quercus robur

Description: The most widespread and usually commonest of several closely related and similar oaks. It is a large, slow-growing tree reaching a height of 45 m or more, characteristic of much of the ancient forests that once covered Britain. The leaves are oblong with lobed edges, and the flowers rather small, yellowish green sprays. The distinctive nuts are known as acorns. It is usually grown as a specimen tree, or allowed to grow out of hedges, but it also makes a good hedge tree, retaining its leaves when it is clipped.

Food value: The true value of an oak is the wide range of insects that live on it and provide food for small birds. A fairly good source of nutrition, the energy value of the acorns is around 100 calories/100 g. They have been used to make flour for human consumption, but the high level of tannins makes it rather bitter. Coarsely ground acorns make a good, cheap bird food.

Birds: The acorns are a particular favourite of Jays, but woodpeckers and a few other species also take them. Jays become very bold in autumn as they move through woods and along hedges in search of oaks and their acorns.

Stone pine

Pinus pinea

Description: The stone pine grows to 25 m, and is also known as the umbrella pine because of its shape. Although the pine nuts sold for human food are generally those from the stone pine of the Mediterranean region, other species are also edible, and are sometimes marketed, including *P. cembra, P. sibirica, P. cemboides,* and *P. gerardiana.* These can all be grown in Britain, and their kernels can be harvested. The mature cones of the stone pine, which are 10–15 cm across, are gathered and warmed to make them expand and give up their seed, which is encased in a hard shell.

Food value: They are rich in carbohydrates and, unusually for nuts, relatively high in vitamin C and thiamin.

Birds: Crossbills are among the few birds able to extract pine nuts from the cones when they are growing on a tree. However, many species will feed on the kernels, including tits and woodpeckers.

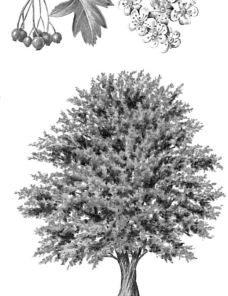

Hawthorn, May
Crataegus monogyna

Description: A large deciduous shrub, or small tree, of the rose family, particularly good as a hedge. Has attractive pink or white flowers in late spring (hence alternative name of May), and a rich crop of berries in autumn. The flowers (and berries) are produced on the previous year's growth, so it is important to not trim all parts of a hedge every year.

Food value: The berries (known as haws) are an important food for a wide range of birds, throughout the autumn and early winter months. The berries are rich in flavonids, essential oils, fruit acids, and have long been used in herbal medicine for reducing blood pressure in humans.

Birds: Thrushes, particularly migrants from Scandinavia often in mixed flocks of Fieldfares and

Redwings, with Mistle Thrushes, Song Thrushes and Blackbirds, move along hedges stripping them of their berries. Many other birds, including Pheasants, also eat the berries.

Barberry
Berberis darwinii

Description: A large group of shrubs, the barberries and their close relatives, the mahonias, are popular as hedging plants, as they are dense, evergreen with prickly holly-like leaves. The Barberry grows to about 3 m high and 3 m in diameter if left unpruned. In spring it has long fronds of yellow or yellowish orange flowers, which later produce blackish blue berries The Barberry (B. vulgaris) also grows wild, and the wild species is often grown as a hedging plant, as it is extremely thorny.

Food value: High in vitamin C, the berries also contain pectins, tannins, berberine and other alkaloids, and for that reason are used in herbal medicine. The flowers produce abundant nectar and attract bees and other insects.

Birds: The berries are attractive to many species, and the dense vegetation is often used by nesting Blackbirds, Wrens and other species.

RIGHT: *Bramblings are closely related to Chaffinches and often seen in mixed flocks, in winter. Although rare in gardens, they can be attracted by beech mast and other seeds.*

Strawberry

Fragaria vesca

Description: The wild strawberry is a diminutive form of the cultivated varieties, with small, rose-like flowers, and leaves with serrated edges. They spread by sending out runners which develop a leaf cluster, which then takes root to form a new plant. The fruits are red with seeds on the skin, and similar to the cultivated hybrid varieties, which are larger but otherwise very similar. The most familiar cultivated varieties are all descended from the Scarlet Woodland Strawberry of the eastern USA, *F. virginiana*, which was introduced into England as long ago as the early 1600s.

Food value: A good source of energy, coming from carbohydrates and sugars, and also extremely rich in vitamin C.

Birds: As every gardener knows, birds, particularly Blackbirds and Starlings, like cultivated strawberries very much, but the smaller, wild varieties of strawberries will be taken by a wide range of bird species.

Oregon grape, mahonia

Mahonia aquifolium

Description: Very closely related to the barberries, and superficially similar, with shiny, evergreen, holly-like leaves. They can be planted in very exposed situations, or used as underplanting with trees. Most varieties have very fragrant pale yellow flowers, that appear in winter or very early spring. The berries are grape-like, blackish blue with a paler 'bloom', and produced in summer.

Food value: The berries are sweet when fully ripened, and used to make jams. The high sugar content should make them attractive to birds, but they often seem not to find them so.

Birds: Most berry-eating birds will eat them, although they generally seem to wait until they are fully ripe and have fallen to the ground.

WHAT IS A FRUIT?

The dictionary definition of a fruit is a 'pulpy, edible seed of certain plants, especially those growing above ground, such as apples, oranges, grapes, melons, berries'. A drupe is a fruit which consists of a pulpy, or fibrous exocarp which contains a nut or stone with a kernel. The exocarp is succulent in the plum, cherry, apricot, peach, etc.; dry in the almond; and fibrous in the coconut. So technically many nuts are fruits. But in general usage, fruits are soft-bodied, or at least fleshy, and nuts are hard.

COMPARISON OF ENERGY VALUES OF APPLES AND PEARS

(Clearly these are only averages, since there is considerable variation between varieties.)

Fruit	Calories/100 g
Apple raw	52
Crab apple	76
Pear raw	56
Apple dried	346

SOURCE: www.nutritiondata.com

Apple
Malus pumila

Description: One of the most familiar of fruit trees, believed to have originated in western Asia. A member of the rose family, closely related to the crab apple, its pinkish white flowers bloom in spring, and the fruit ripens in late summer or autumn. There are hundreds of varieties, and new ones are still being developed. Dwarfing rootstocks are used to create smaller trees suitable for urban gardens. 'Family' trees are also available, with several different apple varieties grafted on to a single rootstock. There are three main groups of apple: dessert, cider and cooking apples.

Food value: A good source of energy, having a high sugar content.

Birds: Generally not eaten by birds until they fall from the tree and are beginning to rot. Fieldfares and other thrushes feed on fallen apples in cold weather.

Crab apple
Malus sylvestris

Description: A deciduous tree, belonging to the rose family, growing up to 10 m. It is one of the wild ancestors of the thousands of varieties of cultivated apples and is probably native to northern Europe. It is usually slightly thorny, bears pink flowers in spring, and small, green-yellow, sour apples in late summer. Like other members of the rose family, the blossom attracts numerous insects.

Food value: Birds usually leave the fruits until late in the season, even into winter, when the fruits have started to break down and are more edible. They have a higher energy content than cultivated apples, which although larger, contain more water.

Birds: Starlings and thrushes, Magpies and Jackdaws, particularly in hard weather. Bullfinches often eat the buds of apples.

Mulberry

Morus nigra

Description: Although widespread in most of western Europe, it is probably a native of western Asia. It is a deciduous tree, with its large dark green leaves appearing in late spring or early summer; it is often one of the last trees to come into leaf. It grows to about 10 m high, but is usually twisted and crooked. The blackish berries are highly prized, but of little commercial significance because they are so fragile and difficult to transport. The closely related white mulberry (*M. alba*) was grown for its leaves to feed silkworms, but its fruit is described as insipid.

Food value: A good source of energy with an exceptionally high vitamin C content, as well as traces of magnesium and potassium.

Birds: Many species, including thrushes, Starlings and warblers, will feed on them, and their droppings will be a characteristic purplish red.

Wild cherry, gean

Prunus avium

Description: A deciduous tree, growing up to 25 m tall. It has peeling reddish brown bark, and pointed, oval leaves. In early spring it has profuse blossom, and the fruits are borne in early summer, in loose clusters. The fruits start yellow, then turn red as they mature. A closely related species, also widespread in the wild, is the bird cherry, *P. padus*, which is a smaller tree whose flowers grow in loose spikes, unlike the those of the gean. Both are very similar to the numerous sour and sweet cultivated varieties of cherry. The sweet cherry is a cultivar of the bird cherry, and sweet cherries have a significantly higher sugar content that other varieties. They are ideal trees for a bird-friendly garden, being both attractive and utilitarian.

Food value: The proportion of stone in the fruit is relatively large, but cherries are

very high in vitamin C as well as having a high energy content.

Birds: Starlings, thrushes and other birds invariably strip cherry trees bare of their fruits before they are fully ripened, so popular are they.

Cherry plum, myrobalan

Prunus cerasifera

Description: A small deciduous tree that makes an excellent hedging plant, but can be invasive as it suckers extensively. It is one of the first trees to flower, with small white flowers in late winter or early spring, followed by shiny, bright green leaves. The fruits are small, round, red or purple plums. It is sometimes used as a rootstock for cultivated plums.

Food value: The fleshy plums are high in fruit sugars, have a high calorific value and are rich in vitamins A and C. They have long been used by humans and make an excellent jam.

Birds: The fruits ripen in early summer, rarely survive until the winter migrants arrive, but are eaten by Blackbirds and other native thrushes, as well as Starlings.

Bullace

Prunus insititia

Description: An early flowering plum, with white flowers, flowering later than the myrobalan, and the tree although thorny is less so than the sloe. The small plums are blackish purple, yellowish green or orange-yellow. They are common in some areas as a hedging, and ideal for a bird-friendly garden. There are several other close relatives that provide good food for birds, notably the cultivated plums and gages and damsons, but these are not as successful as hedging.

Food value: High in sugars, particularly late in autumn. When there is a heavy crop they could be frozen to be given to Fieldfares and other birds later in the year.

Birds: Thrushes, Blackbirds, Starlings, Magpies and many other species eat them, so do Fieldfares and Redwings, but the plums have usually disappeared before they arrive.

Common firethorn

Pyracantha coccinea

Description: The common firethorn is a member of the rose family that if left untrimmed can grow to about 2.5 m high and 2.5 m in diameter. It is a dense, evergreen shrub that has clusters of creamy-white flowers (which attract insects) in early summer followed by bunches of berries. The leaves are shiny green and twigs and branches are very thorny. This shrub is best grown against a wall or as a hedge and it provides excellent nesting sites for many bird species. There are numerous cultivated varieties, differing mostly in the colour of the berries.

Food value: it is not particularly popular with birds, which is useful because by the end of the winter it is often one of the few berries left.

Birds: Many berry-eating birds, including Waxwings, will take this species, particularly during very cold weather.

BELOW: *A Redwing feeding on* Pyracantha *berries. The plant makes an excellent, dense thorny hedge, with the added bonus of berries, in a range of bright colours.*

Pear

Pyrus communis

Description: A deciduous tree similar to the apple, but often significantly larger, and often living to a much greater age. Like apples, pears produce profuse blossom that attracts insects. Their fruit is normally 'pear-shaped', but some varieties are round. Wild pears are rare in Britain. There are many varieties of pears; some very local ones are considered endangered.

Food value: A high sugar level. A good source of energy, and often higher in sugars than apples.

Birds: Like apples, pears are really best left on the tree to fall in winter, when they will be eaten by Fieldfares and other thrushes.

Gooseberry

Ribes grossularia

Description: Closely related to the wild gooseberry (R. uva-crispa) which is found in woods and other shaded areas. It is a spiny shrub with tri-lobed leaves. The egg-shaped, slightly hairy fruits are greenish yellow, but variable in cultivation, with some reddish, or even purple. The fruits are normally 2.5–5 cm long, but the world record weighed 58 g. If left unpruned, gooseberries will grow into large straggling bushes.

Food value: A good source of energy, very high in vitamin C, they also contain vitamins A and D. Can be purchased deep frozen

Birds: Popular with Blackbirds and other thrushes.

FRUITS

Birds eat an enormous range of fruits, and this is to be expected since most fruits have evolved to be attractive to birds, so that they will eat them and then disperse the seeds. The 'soft fruits' of gardeners include strawberries, raspberries, currants, blackberries and similar fruits. But this cuts across botanical taxonomy, since a large number of very differing fruits are members of the rose family Rosaceae, species as dissimilar as strawberries and apples, plums and blackberries. Interestingly, although the family occurs in the tropics, most of the cultivated varieties are from the temperate regions.

RIGHT: *Starlings in winter have a characteristic heavily spotted plumage, and often join with thrushes to feed on fallen fruit in orchards.*

Redcurrant
Ribes sativum
Description: Both the redcurrant and the closely related blackcurrant (*R. nigrum*) are widespread as wild species. They are perennial shrubs, with a distinctive smell. The whitecurrant is a redcurrant lacking red pigment. Currants can be grown in the garden, and if pruned regularly will produce heavy crops. The fruits grow in bunches on short stems, hanging down. As gardeners know, birds will eat them as soon as they are ripe, so they may need protection if you do not wish to give most of your crop to the birds.
Food value: All currants are a good source of energy, being rich in sugars and carbohydrates.
Birds: Eagerly devoured by Blackbirds and Starlings, and also a wide range of other birds, including warblers.

Blackcurrant
Ribes nigrum
Description: Very similar to the redcurrant, but generally larger and more robust, and the black fruits have a slight bloom.
Food value: Similar energy content to other currants, but with a very high vitamin C content. Fresh they have an energy value of about 63 calories/100 g, but dried currants have a value of 283 calories/100 g. However the latter have lost much of their vitamin C content.
Birds: Most small birds will eat them, and netting is needed if you wish to share the crop.

COMPARISON OF THE SUGARS FOUND IN COMMON FRUITS

	Glucose %	Fructose %	Sucrose %
Apple	1.5	6	3
Apricot	1.6	0.9	3
Banana	5.8	3.8	6.6
Cherry	6.1	5.4	0.2
Grape	7.1	6.8	0.5
Orange	2.5	2.2	3.7

SOURCE: www.geocities.com/perfectapple/sugar_carbohydrates_acid.html

Dog rose
Rosa canina

Description: A deciduous shrub, with long, arching, thorny stems, usually grown in hedges. It is related to the numerous cultivated varieties of rose, and several other wild species. The flowers are pink or white, and the berries bright scarlet red. Some of the cultivated roses have large attractive berries, which are also attractive to birds.

Food value: Rose berries (known as hips), have long been known to be rich in vitamin C, and syrups are made for human consumption. A wide range of birds eat them, including thrushes, but they are not as popular as the 'soft fruits', and hips often persist well into winter. They can be gathered in autumn and stored for winter use.

Birds: Thrushes and other berry-eating birds feed on hips, which are also very popular with mice and voles.

Blackberry, bramble
Rubus fruticosus

Description: A deciduous shrub, which comprises a large number of closely related species. It is generally very thorny, suckers extensively, and is very invasive. However, cultivated varieties are less thorny, have larger fruits, and are less invasive. The fruits consist of single-seeded segments, starting green, ripening through red to dark purple. Dewberries are very similar, but with fewer segments to the fruits.

Food value: The fruits are very popular with a wide range of birds, and high (8%) in fruit sugars. The flowers produce copious nectar and attract numerous insects. They (and related fruits) can be purchased deep frozen throughout the year.

Birds: The fruits are popular with many bird species, including Robins, Blackbirds and other thrushes. They are particularly important because of their high energy content for migrating warblers such as Blackcaps and Whitethroats.

FRUIT BUSHES

Soft fruits, such as blackberries and raspberries, as well as currants and gooseberries, together with their countless hybrids and cultivars, are popular with gardeners. However, they are also even more popular with birds (and mice, voles and even foxes). As well as attracting a wide range of wildlife – you may get to share the crop. The most prolific are probably the blackberries, and these will grow particularly well if trained along a south-facing wall or fence. Prune hard in autumn, so that you do not disturb nesting birds in spring.

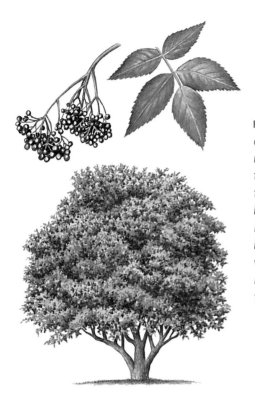

RIGHT: *The elder is one of our most familiar native shrubs and trees. It is a veritable treasure-house for birds, providing luscious purplish-black berries in the autumn, which attract a whole range of birds, such as this Blue Tit.*

Elder

Sambucus nigra

Description: A small deciduous tree, rarely reaching higher than 10 m, with rather brittle, hollow branches. Very fast-growing. In spring it has masses of creamy-white flowers, and in late summer, abundant clusters of small, dark purple berries. It is common in hedgerows and on waste ground, and can be invasive in gardens. There are several cultivated varieties available, and these are generally less invasive. The dwarf elder (*S. ebulus*), is a closely related perennial herb with similar berries.

Food value: Rich in vitamin C, tannins and fruit sugars with high energy value; the berries have also been used extensively for making wines. In spring, the flowers attract lots of different insects, including small beetles, which in turn attract small birds. The berries can be deep frozen, or dried, and put out for birds in winter.

Birds: The berries are popular with a wide range of birds, including Blackbirds, Starlings, Whitethroats and other warblers and Robins. The high (73 calories/100 g) energy values help birds build up fat reserves in late summer, in preparation for migration.

Grape vine

Vitis vinifera

Description: A hardy vine, that has large lobed leaves, it produces small, rather insignificant flowers in summer followed by bunches of grapes. There are literally hundreds of cultivated varieties grown in almost all temperate parts of the world for eating and wine-making. The two main colour forms of the fruit are black grapes (which when dried produce raisins) and white grapes (which produce sultanas). Grapes need extensive pruning each year, and can grow several metres a year. They are best grown on walls, or they can be allowed to climb in trees or on pergolas.

Food value: Very high in sugars, and a good source of energy, with 67 calories/100 g. Grapes contain high proportions of vitamin A thiamin (B1), riboflavin (B2) and vitamin C, and they also contain minerals including calcium, phosphorus, iron and potassium. Raisins and sultanas are more concentrated, since grapes are over 80% water. Raisins have 300 calories/100 g and are about 60% sugar. (The above figures are very approximate, as there is so much variation between varieties.)

Birds: Very attractive to birds such as Blackbirds and Starlings. A single vine can yield at least five or more kilograms of fruit, but birds can strip the grapes very fast once they realize the fruit is ripe.

OTHER TREES AND SHRUBS THAT PROVIDE USEFUL BIRD FOODS

Vernacular name	Scientific Name	Features
Silver Birch	*Betula pendula*	Provides food for insect-eating small birds, including tits and warblers.
Holly	*Ilex aquifolium*	Berries for thrushes in late winter.
Yew	*Taxus baccata*	Berries are eaten by thrushes, paricularly Mistle Thrush, and Wrens and crests feed on spiders and insects among foliage.
Willows & Sallows	*Salix* spp.	Warblers and other species feed on the numerous insects, including caterpillars.
Beets	*Beta* spp.	Any that bolt, the seed can be used for birds.
Olives	*Olea europea*	Ripe (black) olives are favoured by Starlings, thrushes and other birds, but those sold in UK (and elsewhere) are invariably pickled.
Blueberry	*Vaccinium corymbosum*	Rarely grown in gardens, but common in supermarkets.
Cranberry	*Vaccinium macrocarpon*	Exceptionally high vitamin C content. Dried cranberries (usually sweetened) are very nutritious, but have less vitamin C.
True Service Tree	*Sorbus domestica*	An unusual native tree, with small, sour apple-like fruits that are sweet after frost (bletted).
Medlar	*Mespilus germanica*	An attractive tree, which provides food in winter, after the fruits have bletted.
Rowan	*Sorbus aucuparia*	Fast-growing tree with bright orange berries.
Lime trees	*Tila* spp.	Attract lots of insects.
Blackthorn, Sloe	*Prunus spinosus*	Related to bullace, but very sour fruits eaten by birds.
Spindle	*Euonymus europaeus*	Shrub, producing bright pink seed pods eaten by finches.
Cotoneaster	*Cotoneaster* spp.	Produces berries that attact thrushes and Waxwings.
Forsythia	*Forsythia × intermedia*	Attracts Bullfinches.

Live foods

These are perhaps the most difficult to provide for wild birds. However, many of the suppliers of wild bird foods now provide a mail order service, so it is possible to have regular supplies, even if you do not want to breed them yourself. If you have a garden large enough, you can provide supplies of live foods by putting the odd road fatality under a hedge, or leaving a heap of rotting fruit to encourage fruit flies (*Drosophila*).

Ants' eggs

Formica spp.
Description: These are a good live food substitute and are sold in aquarium shops as fish food.
Food value: Probably good, but apparently unmeasured.
Birds: Robins, Dunnock Green Woodpecker. Birds also use live ants in a strange behaviour known as 'anting'. The bird spreads a live ant over its body, or dust bathes in an ant hill. It is believed that this induces the ants to squirt formic acid, which acts as an insecticide on the birds' parasites.

RIGHT: *In late summer flying ants emerge from the nest, and birds such as gulls and Starlings gather to feed on them, providing an interesting aerial spectacle as they swoop and dive on the almost invisible insects.*

Earthworms

Lumbricus spp.
Description: Legless invertebrates, with cylindrical, segmented bodies; each individual contains both male and female organs. They are fossorial, and generally nocturnal, emerging from the ground at night to feed on decaying leaves and other matter, and play a major role in creating humus. Their tail is adapted to anchoring them in their burrows when captured by a bird. Encourage them by keeping a patch in the garden well watered and well mulched with tree bark.
Food value: An important source of protein for many birds.

Birds: Thrushes (including Blackbirds and Robins) and many other birds feed on earthworms, often listening for them after rain.

117

Snails and slugs
Mollusca

Description: Snails are hard-shelled invertebrates, ranging in size from almost microscopic to the Roman snail, which is the size of a plum; slugs lack the exterior

shell. Although generally treated as garden pests, the many different species of molluscs have a wide range of feeding habits and many feed on decaying vegetable matter, while some are predatory. However, seedlings in particular are often prone to damage by snails and slugs. Some slugs exude a slime to deter predators. They can be encouraged in gardens by ensuring hiding spaces are available, such as old brick piles (the mortar is also lime-rich, which snails need for shell formation).

Food value: Widely eaten by humans, some molluscs have provided a staple diet for hunter-gatherers. The larger terrestrial species are often considered a delicacy.

Birds: Most species of slug or snail, particularly the smaller species, are eaten by birds. Song Thrushes are specialist snail predators using an anvil (a stone or other hard object) to smash larger species. An important source of calcium.

Blowfly maggots, gentles
Musca spp.

Description: Small, white legless maggots. They pupate and turn into blowflies. Maggots are sold as gentles in angling shops and are suitable as a live food for birds. Some pet shops also sell crickets and other insects as food for captive reptiles and birds, and many of these are suitable for feeding birds

Food value: A good source of protein and fat.

Birds: Ideal for a wide range of birds, particularly during the summer months, when even normally granivorous birds take mostly insects.

LEFT: *A Song Thrush smashing a snail against a stone 'anvil'. Scattered around the anvil will usually be the remains of numerous snails. Inexperienced young Song Thrushes have been seen smashing cherries in the same way.*

Mealworms

Tenebrio molitor

Description: Small, golden brown caterpillar-like grub, with a shiny skin. It occurs naturally as a pest of stored foods, and also can be found in old nests. Mealworms can be purchased from bird food suppliers and pet shops. They can also be easily bred at home. This is best carried out in an aquarium, glass or plastic, with a secure lid, with only very fine ventilation holes. This container should be filled with sheets of newspaper, interspersed with liberal sprinklings of flour and bran. Then add some mealworms (normally sold by the ounce or gramme), and leave in an airing cupboard or somewhere else warm, adding a slice of bread or apple occasionally, and removing anything that shows the slightest sign of mould. After a few months you will have a flourishing colony of mealworms, and from now on, ensure that the largest ones are always being removed and fed to the birds before they pupate into beetles.

Food value: High in proteins and fats.

Birds: Robins, tits and other small bird relish mealworms.

BARON VON BERLEPSCH'S RECIPE FOR A FOOD TREE FOR INSECT-EATING BIRDS

Ingredients	g
White bread, dried and ground	126
Meat (dried and ground)	84
Hemp	168
Crushed hemp	84
Maw	84
Poppy flour	42
Millet (white)	84
Oats	42
Dried elderberries	42
Sunflower seeds (hearts)	42
Ants' eggs	42

To the total quantity of the dry food as above add about one-and-a-half times as much fat, beef or mutton suet. As the fat easily evaporates in a fluid condition, more suet must be added after the mixture has been warmed several times. It is by no means necessary to keep closely to the recipe; it is only to serve as a guide, and can be altered. The chief part of the mixture must, however, always consist of hemp, whole or crushed.

This mixture is heated on the fire – if prepared at home the fat must be melted and then the dry foodstuff put in – well stirred, and, when boiling poured on the branches of the tree.

All kinds of coniferous trees, especially firs, or else separate branches, can be used to form these food trees. Attention must be drawn to the fact that living trees lose their leaves when hot fluids are poured over them, look ugly, and become easily diseased.

LEFT: *Baron Von Berlepsch coating his feeding tree. A 1907 German complete with moustache! The baron also invented several of the feeder types still in use.*

Other foods

For every food that humans eat, there is probably a bird that shares that taste, so household scraps can provide a useful source of sustenance. Vegetable matter not eaten by birds can be added to the compost heap, which in turn will provide a rich source of worms and other invertebrates. If you live in the country, a visit to an agricultural supplier is a good source of bulk feeds for a large garden. And don't forget that a dish of grit and oyster shell is often appreciated.

Birds: Ideal for thrushes, Starlings and other fruit-eating birds when fresh fruit is not available. As some sorts are liable to swell in the bird's gut, it should be soaked prior to being fed. If finely chopped it will often be taken by Robins and other insectivorous species.

Avicultural and poultry foods

If you have a large garden and want to feed birds, you should certainly consider getting food from an agricultural supplier. It is possible to get organic supplies (obviously organic poultry farms have to be able to obtain them). In addition to wheat, barley and maize and other cereals, there are available pellets particularly suited to ducks and mixtures specifically for pigeons and doves. Some fish foods may be suitable for ducks on larger ponds and lakes, but check the ingredients carefully, and if necessary ask the manufacturers. The advantage of fish pellets is that they are manufactured to float on the surface, so you can avoid overfeeding, and can use them for diving ducks, which normally feed mostly on small animal prey.

Mixed dried fruit

Description: Mixed dried fruit as sold in supermarkets usually includes the following: raisins, sultanas and currants.
Food value: High in sugar content and a good source of energy.

THE RELATIVE NUTRITIONAL VALUE OF DIFFERENT TYPES OF COMMON DRIED FRUITS

Dried fruits	Calories/100 g	Total Carbohydrate %	Sugar %
Apricots	241	63	53
Figs	249	64	48
Bananas	346	88	47
Prunes	240	64	38
Currants	283	74	67
Apples	346	94	81
Sweetened cranberries	308	82	65
Raisins	299	79	59

Oystershell, grit

Birds do not chew their food, but grind it in their gizzard, and in order to do this they need grit. In the breeding season the female has a need for extra calcium to balance that used in egg laying. Grits and calcium in the form of oyster shell or crushed eggshells can be mixed with bird food or supplied in a separate dish. But as anyone who has ever kept chickens will know, they do seem to get through a remarkable amount.

Household scraps

Food value: Cheese is very high in protein and fat, so an excellent food, particularly in winter. The energy values of different cheeses vary widely – grated Parmesan has around 430 calories/100 g; Cheddar has about 400 calories/ 100 g; and Brie and Edam only have around 350 calories/100 g. All cheese is rich in calcium and vitamin A.

Bread, bread and cake crumbs, left-over cooked pasta, rice and other cereals are also excellent foods, high in carbohydrates, and will be taken by a wide range of birds.

Bones hung up will be used by some birds including tits,

BELOW: *Suet pressed into crevices in bark, or in holes drilled in a small log, attracts Great Spotted Woodpeckers and many other species of birds.*

and if larger bones are broken, the birds will extract the marrow.

Fats are particularly important during the winter, but make sure they are well cooked.

Left-over pet foods, particularly cat food, is also highly nutritious, relished by Robins, Blackbirds and Dunnocks.

Hard-boiled egg, chopped up fine, is rich in protein and fat, and also has vitamins A and B12, riboflavin and phosphorus.

Recommended reading

Books

Bill Oddie's Birds of Britain and Ireland
Bill Oddie (New Holland)

Bill Oddie's Introduction to Birdwatching
by Bill Oddie (New Holland)

Birds by Behaviour
Dominic Couzens (HarperCollins)

Birds of Europe
Lars Jonsson (A&C Black)

Birdwatcher's Pocket Field Guide
Mark Golley (New Holland)

Collins Bird Guide
Lars Svensson, Peter Grant, Killian Mullarney
& Dan Zetterström (HarperCollins)

*Handbook of Bird Identification for Europe
and the Western Palearctic*
Mark Beaman and Steve Madge (Helm)

How to Birdwatch
Stephen Moss (New Holland)

*Pocket Guide to the Birds of Britain
and North-West Europe*
Chris Kightley, Steve Madge and Dave Nurney
(Pica Press)

RSPB Handbook of British Birds
Peter Holden and Tim Cleeves (Helm)

Understanding Bird Behaviour
Stephen Moss (New Holland)

Where to Watch Birds in Britain and Ireland
David Tipling (New Holland)

Magazines and Journals

Bird Watching
Available monthly from newsagents,
or by subscription from Emap Active Ltd
Bretton Court, Bretton
Peterborough PE3 8DZ
Tel: 0845 601 1356
emap@subscription.co.uk

Birding World
Available by subscription from Stonerunner
Coast Road, Cley next the Sea
Holt
Norfolk NR25 7RZ
Tel: 01263 741 139
sales@birdingworld.co.uk
www.birdingworld.co.uk

Birdwatch
Available monthly from newsagents or by
subscription from Warners
West Street, Bourne
Lincolnshire PE10 9PH
Tel: 01778 392 027
subscriptions@birdwatch.co.uk
www.birdwatch.co.uk

British Birds
Available by subscription from The Banks
Mountfield, Robertsbridge
East Sussex TN32 5JY
Tel: 01580 882 039
subscriptions@britishbirds.co.uk
www.britishbirds.co.uk

Dutch Birding
Available by subscription from Dutch Birding
Association, c/o Jeannette Admiraal
Lepenlaan 11
1901 ST Castricum
Netherlands
circulation@dutchbirding.nl
www.dutchbirding.nl

Alula
Available by subscription from Alula Oy
Eestinkalliontie 16 D
FIN-02280 Espoo
Finland
antero.topp@alula.fi
www.alula.fi

The Birdwatcher's Yearbook
Published annually by Buckingham Press
55 Thorpe Park Road
Peterborough PE3 6LJ
Tel: 01733 561 739
buck.press@btinternet.com

Useful addresses

Birdlife International
Wellbrook Court
Girton Road
Cambridge CB3 0NA
Tel: 01223 277 318
birdlife@birdlife.org
www.birdlife.org

British Birdwatching Fair
Birdfair Office
Fishponds Cottage
Hambleton Road
Oakham
Rutland LE15 8AB
Tel: 01572 771 079
bbwf@rutlandwater.clara.net
www.birdfair.org.uk

BTO (British Trust for Ornithology)
The Nunnery
Thetford
Norfolk IP24 2PU
Tel: 01842 750 050
info@bto.org
www.bto.org

C.J Wildbird Foods Ltd
The Rea, Upton Magna
Shrewsbury
Shropshire SY4 4UR
Tel: 0800 731 2820
www.birdfood.co.uk

Ernest Charles
Freepost
Copplestone
Crediton
Devon
EX17 2YZ
0800 7316 770
sales@ernest-charles.com
www.ernest-charles.com

Garden Bird Supplies Ltd
Kelvedon Park, London Road
Witham, Essex CM8 3HB
Tel: 0870 899 8989
customerservices@gardenbird.com
www.gardenbird.com

Jacobi Jayne & Co
Wealden Forest Park, Herne Common
Herne Bay, Kent CT6 7LQ
Tel: 01227 714 314
orders@livingwithbirds.com
www.jacobijayne.com

Rob Harvey Specialist Feeds
Kookaburra House, Gravel Hill Road
Holt Pound, Farnham
Surrey GU10 4LG
Tel 01420 239 86
rob@robharvey.com
www.robharvey.com

RSPB
(Royal Society for the Protection of Birds)
The Lodge
Sandy
Bedfordshire
SG19 2DL
Tel: 01767 680 551
info@rspb.org.uk
www.rspb.org.uk

The Wildlife Trusts
The Kiln
Waterside
Mather Road
Newark
Nottinghamshire
NG24 1WT
Tel: 0870 036 7711
enquiry@wildlifetrusts.org
www.wildlifetrusts.org

WWT (The Wildfowl & Wetlands Trust)
Slimbridge
Gloucestershire
GL2 7BT
Tel: 01453 891 900
enquiries@wwt.org.uk
www.wwt.org.uk

Recommended Birding Websites
www.surfbirds.com
www.birdforum.net
www.fatbirder.com
www.birdguides.com

Index

Acknowledgements

Thanks are due to the editorial staff at New Holland, in particular Jo Hemmings, Camilla MacWhannell and Gareth Jones, and to Sylvia Sullivan for many helpful suggestions and comments. I am also grateful to Jacobi Jayne & Co, who supplied much of the food and several of the feeders I used, Garden Bird Supplies Ltd, for supplying foods and a number of images, and Rob Harvey Specialist Feeds, also for supplying foods. My thanks also go to the friends and colleagues, too numerous to name individually, who offered tips and advice.

Photography Credits

All photographs by **Steve Young** apart from:

Alan Marshall: 86 top left & bottom right; 87 top left & right, bottom; 88 top left & right; 89 bottom left & right; 90 top, bottom left; 91 left; 92 centre right, bottom left; 93 top, centre left & right; 94 top; 95 top left & right; 96 bottom left; 98 right; 99 top right, centre left; 100 centre left to right; 101; 118 top left, centre right; 120 right, left; 121 top right & left.

David Tipling: 24 right; 115 top right; 117 bottom left.

Garden Bird Supplies Ltd: 117 bottom right; 119 bottom left.

Artwork Credits

Dave Daly: 3; 10; 19 bottom left; 23 bottom; 25 top, bottom; 30 top right; 36 top; 39 right; 42 top; 43; 45 top, bottom; 46 top, bottom; 47; 48; 59 top right; 64; 72; 74 top; 76 bottom; 79 bottom; 80 bottom left, centre & right; 81 top; 83 top; 84; 86 bottom left; 88 bottom; 89 top; 90 bottom right; 91 bottom right; 92 top; 93 bottom; 98 left; 100 bottom; 117 top.

Richard Allen: 12; 14; 15 right; 17 bottom; 19 bottom right; 21; 22; 26; 28; 33; 35 right; 37; 39 left; 41 left; 42 bottom; 56 top, bottom; 61; 66 bottom; 68 top; 71 bottom; 78 bottom; 91 top right; 94 bottom; 99 bottom right; 103; 118 bottom left.

Wendy Brammall: 27.

Wildlife Art Ltd: 67; 102; 104 top, bottom; 105 top, bottom; 106 top, bottom; 107 top left & right; 108 top, bottom; 109 top, bottom; 110 top right & left, bottom; 111 top, bottom left; 112 top right, bottom left; 113 left, right; 114 left, right; 115 top left, bottom right.